Management decisions for engineers

James Parkin

Thomas Telford

Published by Thomas Telford Publishing, Thomas Telford Services Ltd, 1 Heron Quay, London E14 4JD

First published 1996

Distributors for Thomas Telford books are
USA: American Society of Civil Engineers, Publications Sales Department, 345 East 47th Street, New York, NY 10017-2398
Japan: Maruzen Co. Ltd, Book Department, 3–10 Nihonbashi 2-chome, Chuo-ku, Tokyo 103
Australia: DA Books and Journals, 648 Whitehorse Road, Mitcham 3132, Victoria

A catalogue record for this book is available from the British Library

ISBN: 0 7277 2501 7

Typeset by Techset Composition Limited, Salisbury, Wiltshire
Printed in Great Britain by The Lavenham Press Limited, Suffolk

To

Khun Cheewan Narangajavana

Preface

Decision making is at the heart of managing. It is the retraceable expression of the process of organizing. Good decisions produce good outcomes for both the organization and the manager, and careers may be made or broken by a single decision. This book is a guide, designed for engineering managers, to the theory and practice of how good decisions are made and why poor decisions occur.

Engineers take for granted the decision making methods of design, construction and production. Guided by sign posts such as codes, specifications and drawings, engineering is intrinsically tied to the formulation and interpretation of rules for decision making. Our medium of work is matter, which is, to a large degree, predictable. We have learnt that to create a bridge or an engine, we must follow certain design rules to create a set of documents that will closely determine the decisions that a contractor will make to build the object. In practice, we operationalize physical laws and transform them into decision tools.

But the management of people and organizations is radically unlike normal engineering processes because human beings are not predictable. They insist on making up their own minds which they change at will; they persist in a rationality unlike that of science, and they respond to manipulation by modifying the rules. Above all they are complex and unknowable. This otherwise happy situation creates problems for the engineer who enters a management role. He or she must learn new tricks, at least in part, from books such as this.

This book is the result of many years of teaching and consultancy in organizational and social decision making. It is an expression of a knowledge base which draws on some 30 years of civil engineering practice. Nevertheless, it cannot provide the certainties of science. What it does try to do, however, is to gather in a reasonably short and accessible form, some of the most important concepts in managerial decision theory and to make some suggestions as to how they can be used to help the new, and perhaps not so new, engineering manager.

I have not assumed that the reader will be familiar with general management theory and have attempted to set the scene with a review of the salient management concepts. After that the reader will be guided through what psychological, mathematical and sociological research has discovered about individual and organizational decision making. Wherever possible, this knowledge has been distilled into practical guidelines for action.

Decisions concerning the form and contents of this book have been made easier by others. In particular, I wish to thank Sally Smith of Thomas Telford for her editorial guidance, the referees for their perception, and Dr Ray Cooksey and Dr Deepak Sharma for their comments on early drafts. As in all decision processes, sometimes their inputs were gratefully received but not wholly adopted. Finally, my thanks to my wife Ankna, whose patience and word processing proved critical to timely completion.

Permission to adapt copyright material is gratefully acknowledged as follows: *Administrative Science Quarterly*, © 1976, for Fig. 15; Avebury Publishers for Fig. 19; Blackwell Publishers, *Econometrica*, © 1979, for Fig. 6; Dr Ray Cooksey and Academic Press for Fig. 14 from *Judgement Analysis: Theory, Methods and Applications*, © 1995; P. Dryden and L. Gaweki for Figs 10 and 11; Elsevier Science Ltd, *Geoforum*, © 1994, for Fig. 16; Metrica, Inc. (sponsored by the US Air Force) for Figs 7, 8, and Tables 5–8; Walter de Gruyter and Co, *Organizational Studies*, © 1989, for Table 3.

Jim Parkin
Sydney

Contents

Part I

The management setting

This opening section of the book examines the nature of management and its significance to the decision maker. Chapter 1 describes what is known about the management function and how managers behave, think and communicate. It reveals the complex and dynamic character of management work and the importance of human interactions, persuasion, conflict and power. It indicates the necessity to understand decision making, and to frame recommendations concerning decision management, in ways that do not rely on an abundance of time for calm contemplation.

Chapter 2 reviews the nature of public and private organizations within which most managers work. Organizations are ubiquitous and infinitely varied in function, so generalization is dangerous. Nevertheless, it may be said with some confidence that all large organizations can be explained as an interaction of structural factors, built around hierarchy, rules and norms of behaviour, with the dynamics of individual interests, conflict and power. Some organizations may best be described by emphasizing the structural factors and others by highlighting the results of individual actions. Nevertheless, all will contain elements of both modes of ordering. It is concluded, therefore, that decision making will be moulded by these tendencies and expressed as rule following or problem solving methodologies.

Chapter 3 discusses the problems experienced by engineers when they find themselves taking on an increasingly managerial role. Some of these problems are personal and the product of a way of thinking about situations common to all professionals. This cool linear rationality sits uncomfortably with the messy world of management and feeds an unrealistic expectation that the logic of engineering can be applied to organizational decision making. This natural personal disadvantage may combine with those problems associated with exclusion from the top ranks of corporate leaders. However, the world of the manager is changing quickly, and in as many ways as the rest of modern society. Technical expertise may well be the key to effective organizational decision making in the coming decades.

1

Management

What are managers for?

Those that do the organizing in organizations are often called managers. Indeed, today, practically everybody that does not perform a purely technical role, or is not in the bottom layer of the hierarchy, may claim to be a manager of some sort. Commonly used definitions of managing such as 'getting things done through people' (Babcock, 1991) imply that, provided you have a subordinate, then you manage. Certainly at this level, many of the skills required of a manager start to be exercised, and it is the necessity to direct people rather than concepts or machines that first exposes the engineer to the realization that engineering science alone will not get the job done. The engineer is faced for the first time with the curious notion that other people must be made subject to his or her will if the project is to be completed. However, although the manipulation of people and technology (rather than technology alone) is the most important role that distinguishes the engineering-manager from the engineer, it is, nevertheless, an inadequate description of management. A more comprehensive definition must be related to the process of organizing.

A useful definition that should appeal to engineers, and certainly links the manager to the organization, has been provided by Whitley (1989). Building on the work of Penrose (1980), he claims that

> the basic function of management [is identified] as the construction, maintenance and improvement of an administrative system which co-ordinated and transformed human and material resources into productive services (p. 211).

Although this definition is somewhat mechanical in character, it serves to illustrate the difference between management and administration, as well as that between engineering and management. In particular, it is the discretion required of a manager—the ability and necessity to make decisions—that sets the manager apart from the rule-following administrator. This discretion is, of course, constrained by rules and the collective nature of organizing but is, nevertheless, a common thread that runs through the work of any manager. The organizational nature of management, and its many functions not evident from the authority structure, is also reflected in this definition. Those in the hierarchy in charge of substantial divisions doing routine work may not have much of a decision making role, whereas some key personnel without many subordinates may influence the flow of resources across the organization on a daily basis. Thus, the degree to which a person manages rather than administers is manifest in their agency power to make resource decisions.

Whitley points out that this definition implies that management has a number of task characteristics which differentiate it from other jobs in organizations. Firstly, the task content is highly dependent on the nature of the resources being managed—the management of an engineering firm is not the same function as that of a department store manager. Secondly, as management tasks are contextual and contingent they are resistant to standardization and procedural formalization—tasks change with time and circumstances. Thirdly, management tasks involve the tensions and conflicts resulting from the continuing development and restructuring of the organization and the conflicting requirement for stability and control. Lastly, the effectiveness of management tasks can only be judged collectively, if at all.

However much we may speculate about the nature of the tasks required of a manager, we cannot necessarily predict much from them about the behaviour of real managers. This is not self evident, however, and classical management writers appeared to assume a strong connection. Perhaps up until the early 1970s most managers, when asked what they did, would reply with the list of tasks derived from Henri Fayol's work in the 1920s and 1930s—they performed planning, organizing, leading and controlling tasks (POLC). These terms had become part of the ideology of managerialism (Alvesson, 1987) which conceived management as a context-free professional pursuit. Their apparent rationality falls short of the task-dependent, contextural, contingent, conflict ridden and collective nature of real management tasks. Whether or not the POLC model is a good description of the management function, it is clearly a poor description of what real managers do at work.

What do managers do?

Perhaps the first, and certainly the most important, reaction to the POLC model of management came from Henry Mintzberg's (1973) study of the work of five chief executive officers. These gentlemen did not spend much time reflecting and planning but rather their time was taken up with frantic activity during long and fragmented days. Three main types of roles could be detected in all this activity. Firstly, interpersonal roles which came from their position of leadership and the need to liaise with outsiders. Secondly, information roles involving the collection and dissemination of information of various sorts. Lastly, the decisional roles involving the initiation of projects, handling conflict, allocating resources and negotiating.

Mintzberg opened a floodgate of empirical work on the behaviour of managers at all levels in the organizational hierarchy. After a systematic review of the results of this research Hales (1986) concluded that management work is

- contingent upon, *inter alia*: function, level, organization (type, structure, size) and environment (p. 100)
- sufficiently loosely defined to be highly negotiable and susceptible to choice of both style and content (p. 101)
- not part of a neat, coherent, unproblematic set. Activities may be competing, even contradictory, and this itself produces the important managerial work of compromise and negotiation (p. 102)
- an unreflective response to circumstances (p. 102)
- very different from how it is reputed to be (p. 103).

He summarized the features of managerial work recalled by the empirical evidence as follows (p. 104)

- It combines a specialist/professional element and a general, managerial element.
- The substantive elements involve, essentially, liaison, man-management and responsibility for a work process, beneath which are subsumed more detailed work elements.
- The character of work elements varies by duration, time span, recurrence, unexpectedness and source.
- Much time is spent in day-to-day trouble-shooting and *ad hoc* problems of organization and regulation.
- Much managerial activity consists of asking or persuading others to do things, involving the manager in face-to-face verbal communication of limited duration.
- Patterns of communication vary in terms of what the communication is about and with whom the communication is made.
- Little time is spent on any one particular activity and, in particular, on the conscious, systematic formulation of plans. Planning and decision making tend to take place in the course of other activity.
- Managers spend a lot of time accounting for and explaining what they do, in informal relationships and in politicking.

- Managerial activities are riven by contradictions, cross-pressures and conflicts. Much managerial work involves coping with and reconciling social and technical conflict.
- There is considerable choice in terms of *what* is done and *how*: part of managerial work is setting the boundaries of and negotiating that work itself.

It is not possible to leave the question of what managers do without mentioning the investigation of 457 managers by Luthans *et al.* (1988). They identified four main groups of activity

1. *routine communication*—exchanging information and doing paperwork
2. *traditional management*—planning, decision making and controlling
3. *networking*—interaction with others and socializing/politicking
4. *human resource management*—motivating, disciplining, conflict management, staff selection and training.

The researchers were interested in the relationship between the time spent on these activities and measures of success. For comparison they used about half the sample as a control group. These managers, on average, spent about 30 per cent of their time on each of the communication and traditional management functions, and the remaining 40 per cent evenly divided between networking and human resource management. Luthans *et al.* (1988) then isolated 178 managers who, on the basis of unit performance and subordinate satisfaction, were considered effective. This group spent considerably more time on routine communications, more on human resource questions and much less on traditional functions and networking. The final group were those managers that had achieved rapid promotion. It will not surprise experienced managers to find that these 52 individuals spent on average, 48 per cent of their time networking and about the same as the control group on routine communication. Only about one quarter of their time was spent on traditional management functions and human resource management. I will refrain from stating the obvious.

Interesting as these empirical results are, they must be put in context. Most of them are from the USA and may not be generalizable to other cultures. Nevertheless, the work of Hofstede (1984) indicates that English-speaking countries such as the UK and Australia share a very similar set of organizational values with the USA, which would imply some commonality in approach to management. A second caveat is in order. The huge range of jobs within the organizational world ensures that management means very different things to different people. Indeed, different people in the same management job will interpret the requirements in quite different ways, and negotiate anew the action boundaries.

What do managers think about?

Isenberg (1988) studied the thought processes of a dozen senior managers in the USA. He found that only very rarely did they think in the linear, rational way that requires the formulation of specific goals, alternatives and the optimization of some function. Rather they focused on 'how to create effective organizational processes and how to deal with one or two overriding concerns, or general goals' (p. 528). The processes that dominated their thinking were largely communicative—building networks within and outside the organization to enable them to improve organizational functions. How to get people together, how to motivate, how to create interdependences and align interests for the good of the whole organization. This involved getting to know the key players and how they thought. This, in turn, required high level skills in communication, listening, persuading and negotiating to be effective. While going about the job of getting others to think about the business, the senior manager would keep in the back of his or her mind a few overriding but quite general goals—how to turn around a subsidiary, increase market share or improve the unit discipline. These concerns would add shape to what otherwise appears to be a set of random interactions.

Both Isenberg (1988) and Mintzberg (1984) agree that senior managers do not appear to use the very rational calculative strategies associated with the left brain lobe. It is the more patterned, intuitive skills of the right lobe that are favoured. This does not imply a lack of rationality. The fact that managers overwhelmingly preferred verbal to written communication can be explained by the 'desire for relational, simultaneous methods of acquiring information' (Mintzberg, 1984, p. 192). Eye contact, body language and real time exchanges are far richer in data than a memo. Thus, the faculty of judgement requiring the simultaneous integration of a wide range of hard and soft data is preferred to analysis. Patterns of data, frequently used, are applied intuitively to make rapid decisions in a fast moving day. This is judgement in action. The patterns of data required for judgement can be seen at another level. Isenberg noted that senior managers put a number of problems into a network, or portfolio, of relations—a problematic data bank for future judgements at the right moment.

Before we are carried away by the image of the active, integrative manager, thinking only of the good of the organization, we should perhaps contemplate another more human feature of a manager's thinking. Managers are not just functioning units of an organization, performing whatever tasks are required to obtain effectiveness, and thinking only of the processes required to achieve that end. They are, above all, individuals with careers, fears, ambitions and faults. They have interests and desires separate from organizational objectives. In addition and above all, they need to survive in a harsh work jungle.

Robert Jackall's (1988) empirical work was not constrained by respect for the ideology of management—what management said about itself to sustain its prestige and influence. His concern was with the moral climate of managing. His results spoke of managers whose daily work-lives were dominated by questions of ambition and fear of failure—managers who were acutely aware of the contingency of their positions. This is a world where image is important. Wearing the right clothes, having 'sound' opinions, laughing at the boss's jokes, appearing relaxed under stress, working long hours and above all, networking, networking, networking. Survival is seen in terms of avoiding blame for your and others' mistakes, of having a higher level protector, of alliances, of passing the credit up, and being a team player. Morally, it is an ugly world of fear and greed. It may seem an exaggeration and no doubt it is, but it is a picture not unlike that painted by Luthans *et al.* (1988) of those managers who were not necessarily effective, but nevertheless, achieved rapid promotion.

It is worth noting that this short section on management thinking is a key to much of what follows in the rest of the book. The nature of judgement will be discussed at length in parts II and III where it is also demonstrated that networks, mediated through communication, are a key feature of organizational decision making.

How do managers communicate?

As we have seen, Luthans *et al.* (1988) found that between half and three quarters of a manager's time was spent communicating in some way with others within the organization, or to suppliers, customers and other interested actors. Much of this communication was verbal; either face-to-face or on the telephone (Hales, 1986). And moreover, much of it was idle talk or gossip. Standing around the coffee machine, leaning on the door jamb, sending jokes through the electronic mail and trading views on the local football team are not designed to enhance the objectives of the organization. Yet they are a thread of enjoyment that is the antithesis of the fearful, ambition-driven image of the manager described by Jackall (1988). These are the incidents that make organizational life bearable for many and zestful for others. Why don't the bosses stamp it out? Well, of course, consciously or unconsciously they realise that the difference between an organization with some degree of cohesion and a mere assemblage of individuals is social interaction of this sort. Through jokes and gossip, the images, language and interpretations of the established members are relished and transmitted to new employees. Cultures are created and transmitted (March and Sevon, 1988). This tribal instinct is so strong that subcultures among some groups of employees may develop that are antithetical to those desired by top management. This is a tough situation to handle because changing these awkward group norms can be difficult. Nevertheless, within a subgroup, it is the jokes and idle chatter that often gives a working day meaning—the meaning that makes getting out of bed just that little bit more bearable.

Moreover, as Luthans *et al.* (1988) has demonstrated, talking to the right people seems to accelerate promotion. Networking is an efficient way of gathering information and feedback which may be useful to a career. It also enables alliances to be built among managers with similar, non-competitive interests. It is not surprising, therefore, that ambitious managers spend considerable time on the telephone talking of things that may be of little immediate consequence to the unit's objectives, but will nevertheless forge another link with someone who may be useful in the future.

Between the gossip and the jokes the work has to be done, and communication has a central place in the way managers organize. In its most obvious and simple guise, communication is the transmission of information up, down and across organizations and to external parties. Thus, data is gathered from outside or inside the organization, and directed to those managers who need it for decision making. Directives, memos and telephone calls come from senior managers for action at lower levels. Good news messages from the chairman are sent to all the troops to nourish morale. Unfortunately, many of these messages do not succeed in their purpose. Messages are distorted as they are reinterpreted at each stop in the transmission chain; unpleasant information may be filtered, and strategic information stopped and hoarded for future use in a power struggle. In an effort to condense a message the full force of an argument may be lost, people may not take seriously a message from a non-credible source, messages that do not conform to the receiver's prior expectations may be ignored, and too much information may overload the manager such that only the beginning and end of a long memo will be retained (Stohl and Redding, 1987).

As management is about judgement, and judgement is improved if a rich mix of cues are used, the nature of the communication will affect its usefulness. *One way* messages without any feedback are the least effective. If the message is distorted in transmission or misinterpreted then the effect may be highly undesirable. What is 'immediate' for a busy manager may be very different for a stock analyst, and a polite but firm instruction may be interpreted as only a suggestion by an insensitive new employee. *Two way* communication has the advantage that feedback can be obtained and, if necessary, the message can be rephrased or repeated. Experienced managers prefer verbal communication for this reason (Mintzberg, 1973). Not only the data can be used as cues for judgement. Body posture, the use of the hands, the look in a person's eye and the confidence of the delivery may be critical to the success of the message. How something is said is often more important than what is said. The credibility of the source can be judged from such symbols as grey hair, plush offices, well-cut suits, old school ties and clan or regional accents. Like all judgement however, the availability of multiple cues that have, in practice, little bearing on the issues can be very misleading. Some research quoted in Mitchell *et al.* (1988) illustrates this admirably. A single person appeared before 24 different interview panels who asked her the same questions each time. In half of her interviews she showed apparent enthusiasm by using good eye contact, smiling and other relevant body signals. In the other interviews she used the body language of aloofness. The results clearly indicated that her scores on likeability, qualifications, competence and likelihood of success were rated very much higher by the interview committees who were given the 'enthusiasm' message. They would have employed her but the others would not. As we will see, when judgement is discussed later, human cognition is prone to many biases of this sort. However, just because the cues were behavioural in this case does not make them intrinsically less reliable than any other. Because a memo (written cues) comes from the managing director does not necessarily mean that the data discussed or the opinions expressed are valid. Nevertheless, this memo is more likely to precipitate action than one containing the same message from the junior manager. Perhaps, in practice, the reason why most personnel hiring is done using both a written message (the cv) and an interview is to combine the advantages of multiple behavioural cues with a set of 'cool' written cues. Certainly, the combination would seem to be a potentially effective way to convey important messages which are meant to influence others.

How do managers influence others?

'Getting things done through people' is central to the management function (Babcock, 1991), and it is inconceivable that the coordination and transformation of 'human and material resources into productive services' (Whitley, 1989, p. 211) could be achieved without the cooperation of a wide range of actors from inside and outside the organization. Thus, management effectiveness is highly dependent on the ability to influence others—one way or another.

The empirical work of Kipnis and Schmidt (1983) provides a convincing description of how managers attempt to influence others. These researchers first asked managers to describe actual influence incidents. The 58 influence tactics that were identified from their work were used on a new sample to determine the frequency of their use to influence their superiors, peers or subordinates. Factor analysis of these results revealed seven strategies of influence, as shown in Table 1.

From these strategies, a measuring instrument called the profile of organizational influence strategies (POIS) was used to explore patterns of their usage in Britain, Australia, and the USA. Using POIS, each manager indicated how frequently they used each strategy to influence superiors and subordinates. The results were

Table 1. Strategies of organizational influence (Kipnis and Schmidt, 1983)

Strategy	Behaviour
Reason	The use of facts and data to support the development of logical argument. Sample tactic: 'I explained the reasons for my request'.
Coalition	The mobilization of other people in the organization. Sample tactic: 'I obtained the support of co-workers to back up my request'.
Ingratiation	The use of impression management, flattery, and the creation of goodwill. Sample tactic: 'I acted very humbly while making the request'.
Bargaining	The use of negotiation through the exchange of benefits or favours. Sample tactic: 'I offered an exchange (if you do this for me, I will do something for you)'.
Assertiveness	The use of a direct and forceful approach. Sample tactic: 'I demanded that he or she do what I requested'.
Higher authority	Gaining the support of higher levels in the organization to backup requests. Sample tactic: 'I obtained the informal support of higher-ups'.
Sanctions	The use of organizationally derived rewards and punishments. Sample tactic: 'I threatened to give him or her an unsatisfactory performance evaluation'.

similar for all three countries. When influencing subordinates, *reason* was the most popular followed by *coalition, ingratiation* and *bargaining,* and *assertiveness* and *higher authority* were the least. Power affected the strategy used. Subordinates tended to use ingratiation more when expectations of a good outcome were dim or the objective was personal. Assertive behaviour was more often used from a position of power and the objective was organizational. Generally, reason was the strategy of first choice in most situations.

It is of further interest to note that this research revealed three clusters of influence use. *Shotgun* managers were inexperienced and ambitious and although they tried many influence strategies, they were relatively unsuccessful. *Tactician* managers were generally in charge of technologically complex work requiring sophisticated skills, could manipulate their own budgets and policy, and had high work satisfaction. These managers successfully used reason far more than any other strategy. *Bystander* managers tended to be long-serving managers in charge of large units doing routine work. They had little influence over budgets, personnel or policy. These managers had the least scores on all seven strategies—they had given up trying to influence others.

What about the unwritten rules?

Much of the behaviour discussed by Jackall (1988) involved following unwritten rules (March, 1994). The dress code, working long hours and the right social responses are learnt by the ambitious manager very soon after joining the organization. He or she will look up at the bosses and around at their peers to learn the appropriate rules associated with being a chief engineer or cost accountant. Sometimes, their role identity is often subsumed under a global identity such as 'decision maker', 'team player' or 'professional', and reinforced (or otherwise) by the organization through appropriate rewards. Often these uncodified rules control the behaviour of a manager as much as the official organization regulations. Other identities are left behind. Yesterday I was an engineer, now I am a manager. Good parent, football player, political activist or country and western dancer are abandoned outside the front door and only mentioned if they reinforce the required role image (that is why so much jogging is done in the lunch hour). So, managerial action is often more related to role symbolism than rational calculation—to the playing of the appropriate managerial game rather than the requirements of the corporate plan.

Where does management power come from?

Power is the ability to act. This definition can be found in just about any major dictionary, but its simplicity hides the true complexity of the use of the term and the associated theoretical issues. The implication of this definition is that power is to do with the *power to* facilitate some action. This is the interpretation shared by Arendt (1970), who considered power to be a group property which emerges from the group's common action. This power can be bestowed on one person by the group—thus, the Prime Minister is given power to govern the country by the people. In a similar way, the sociologist Parsons (1963), associated power with legitimate authority. But power has a dark side in the popular consciousness. This is the *power over* people much associated with force and coercion at one extreme and the manipulation of agendas and lying at the other (Lukes, 1974). This is the power to overcome resistance. This is the assertive and sanction power so reluctantly used by the more senior manager in the previously discussed study on organizational influence. Of course, managers do use such power if called for, but with caution—aware as they will be, of the possible negative effects on organizational cohesion and morale.

In organizations, power is an effect produced by the act of organizing. It is intimately associated with individual and organizational action—it is, in fact, an agency performance (Giddens, 1984). As Crespi (1992) says, power is manifest when an individual or organization takes decisive action in the face of the constraining rules and norms of a social structure. Power is what is seen when managers manage rather than administer. We are, therefore, back to the original definition—power *is* the ability to act.

Where does this ability to act come from? What enables a manager to manage? Morgan (1986, p. 159) lists the following sources

- *Formal authority* is the power given to an individual by their role in the organization and is legitimated by the tacit agreement of those subjected to it.
- *Control over scarce resources.* As the Kipnis and Schmidt (1983) study shows, this is a potent source of power.

However, it is not normally used to gain compliance unless reason fails.

- *Use of organizational structure, rules and regulation.* Control within organizations can be concentrated in particular locations by the adroit manipulation of existing rules or by internal reorganization.
- *Control of decision processes.* What subjects are on the agenda, how the process is to be conducted and who is involved are very much the subject of Part III of this book.
- *Control of knowledge and information* is the power source of professional groups or highly skilled technicians. Many 'political' managers also collect, stop, spread or fabricate information to further their own interests.
- *Control of boundaries.* Groups within organizations may be careful to control the flow of information across its boundaries to restrict the potential for outside control. Secrecy and agenda setting can be used by secretaries and special assistants to control what goes in and what comes out of the chief's office.
- *Ability to cope with uncertainty.* Those that can predict market trends and those specializing in organizational troubleshooting both gain their power from their ability to protect management from the disruption of the unexpected.
- *Control of technology.* The introduction of new technology often changes the balance of power in organizations. The control over data or any of the core functions may shift and with it the power.
- *Interpersonal alliances, networks, and control of informal organizations.* Having a friend on high and a network of first name colleagues in other divisions can enable a manager to take action not available through the official channels.
- *Control of counter organizations.* Although a union official may not have much apparent role authority, his or her

control over the functions of key labour groups may give them significance in the life of the organization.

- *Symbolism and the management of meaning.* One of the achievements of leadership is to get others to share your sense of meaning—to define their realities for them through the use of symbols, such as mission statements, logos, group retreats and ceremonies. Moreover, a manager may seek to enhance personal power by impression management and gamesmanship.
- *Gender and the management of gender relations.* Organizations were usually created by men for men. Procedures and structures are, often unconsciously, created to give men more opportunity for power than women.
- *Structural factors that define the stage of action.* The nature of capitalist society, and its organizations, determines the nature of the deep structures underpinning power. The reflection of general class relations in organizations is an example of this.
- *The power that one already has.* The ability to get things done can stimulate a feeling of empowerment that leads to greater achievements. People can grow into powerful jobs.

These management sources of the ability to act are only indicators of the complexity and richness of real situations. All human actors who take part in organizing draw that ability from the problematic situation that is the organization. Even the most dedicated bureaucrat has considerable discretion in practice, and even the office cleaner makes choices. So, to some degree, all have power. In most cases some of that ability to act will be manifest as a *power to do* something as well as a *power over* other actors. In case of managers, however, their power is supported and nourished by the ideology of management and the social enframement of power sources within the structure of the organization.

Implications for decision making

Management is the use of power; management is decision making; management is contingent and action orientated. Management requires the use of complex communication networks both within the organization and with external interests. Within these networks, decisions are made and action is taken using the medium of language. Conflict, politics, argument and persuasion are the norm. Cool rationality rarely seems to be evident in the decision making process, yet rational persuasion seems to be the way all interests are best brought into alignment. Above all, managers use judgement—judgement based on as rich a mix of cues as can be obtained given the time constraints. No model of organizing can ignore these facts. To exclude considerations of self interest, politics, judgement and persuasion is to distort reality for the sake of purity.

Summary

- Real management behaviour is not adequately described as planning, organizing, leading and controlling. It is, in practice, highly varied and task dependent and an ability to communicate, persuade, handle conflict and think intuitively are essential.
- Rules of conduct and corporate norms are important to managers. However, management requires judgement and decision making which is often unaided by the organizational rules.
- A manager's power to act without the guidance of rules may come from sources other than that of formal authority.
- A manager's life does not contain many moments of calm contemplation and our views about the practice of decision making must take account of this reality.

2

Organizations and management

The importance of a point of view

In the social sciences there are few questions as gripping as the dichotomy between individual agency and social structure—how human beings reconcile their personal identities with their roles as members of social institutions. If a person gains their sense of identity largely from their own inner life they may become alienated from society—a recluse, an outcast or a mental patient. Too much identification with social institutions and the individual consciousness tends to disappear and action outside the group becomes difficult (Crespi, 1992). What is generally accepted depends to some degree on which set of national norms governs behaviour. In many East Asian countries loyalty to the collective is highly valued, but in the USA, UK and Australia a much more individualistic view is taken (Hofstede, 1984). However the relationship is resolved, the struggle for individual or collective identity is at the heart of our attitude towards human institutions.

The social institution that concerns us in this chapter is the organization. Since the rise of industrial society 200 years ago, the life of humanity in modern society has been dominated by the organization. These goal-directed institutions either employ us or, if we have somehow escaped, they dominate the frameworks that make modern life possible. It is the relationship between we human beings and our organizational creations that is our subject, and in the process of the discussion the tense and dichotomous relationship between our sense of individual agency and our role as loyal employees will be examined.

Sociology is the discipline that has shaped much of the theorizing concerning organizations, and sociology has a tendency to place more emphasis on structure than agency. This is not surprising because mainstream sociology was created during the nineteenth century reaction against the revolutionary excesses of the eighteenth century. An urge for control informed the early sociology of institutions (Dawes, 1970). After some mutations, the concept of social system threw up the dominant model of organizations as institutions stabilized by a central value system. This 'systems frame of reference' (Silverman, 1970) was particularly popular during the 1950s and 1960s and has since mutated into the now orthodox and pervasive 'contingency theory' model (Willmott, 1990). It is true that prior to the Second World War, structure was also dominant, but in a less flexible and organic way, with a great emphasis placed on the internal functions of organizations rather than their capacity to adapt to environmental forces which so fascinated the post war researchers. After the war, however, the emphasis has lain with a common set of goals, values or (recently) culture. Human agency is de-emphasized and the organization is treated not as a mere collection of interacting human beings, but rather as a thing with a life of its own. A thing with values.

Reacting against this orthodoxy, Dawes (1970) sees another underground stream of consciousness flowing from the emancipatory ideals of the eighteenth century Enlightenment. This holds to a view that human beings act out of a desire to achieve some ideal and are driven by an urge to mould social institutions to fit that ideal. Thus, they struggle to impose their sense of meaning on an organization. Instead of a tendency to submerge the individual into the collective system for the good of the system, this view sees struggle, conflict and argument dynamically shaping and reshaping the organization. This is the precursor of the body of organizational research that reacted against the systems view in the 1970s and continues to counter the excesses of structuralism. Meaning, identity, interests, conflict, persuasion and games are some of the key words of this point of view. Agency rather than structure.

Analogous dualities will shape this chapter. But rather than treat social structure and individual agency as forces (like good and evil) in opposition, I would rather treat them as a dialogue between aspects of the relationship between human beings and their organizational creations. Creations that are so real and powerful but are, nevertheless, ultimately subject to the human will. As a matter of convenience I will treat those aspects of organizing to do with structure, hierarchy, function, control and adaptability as part of the *administration mode of ordering*, and those concerned with meaning, interests, power, politics and symbolism as part of the *enterprise mode of ordering*. These modes of ordering are derived from the empirical work of Law (1994). They are the two principal patterns, or ways of doing things, which underlie the interactions of the research organization he investigated, and these can readily be seen to be roughly equivalent to the two underlying views of organizations discussed earlier. One emphasizes structure and the other agency. It should be noted that these modes of ordering do not exist alone but, rather, coexist in time and in terms of behaviour and interaction. Moreover, two other less dominant and universal modes of ordering—*vocation and vision*—may also be found in organizations with high numbers of professionals or with a charismatic leader. For our purposes, they will be subsumed within the two dominant modes.

Law (1994) explains that modes of ordering can be inferred from the observed actions and recorded conversations of managers. They are regular patterns that can be observed in the networks of interaction in organizations. In his words,

> recurring patterns embodied within, witnessed by, generated in and reproduced as part of the ordering of human and non-human relations (p. 83).

Although I am in danger of violating Law's particular use of the term, it is interesting to see the similarity between the concepts of modes of ordering and that of organizational culture. Bloor and Dawson (1994) propose a particularly useful definition of organizational culture

> a patterned system of perceptions, meanings, and beliefs about the organization which facilitates sense-making amongst a group of people sharing common experiences and guides individual behaviour at work (p. 276).

To call such a thing a mode of ordering would seem to overcome the misunderstandings associated with the anthropological and artistic uses of the word culture. Bloor and Dawson observe that not just one organizational culture (mode of ordering) exists, but rather many nested or parallel cultures generated by different groups, or within a single group, and activated by different circumstances at different times. With this caution in mind, we will now discuss organizing under the two principal modes of ordering—the administration mode and the enterprise mode.

It is noteworthy that despite the existence of overarching theories such as the systems view, in practice, organizational functions can best be explained in terms of models of what managers do when they are organizing. People cannot be explained away by grand theories.

The administration mode of ordering

In the words of Law (1994):

> Administration tells of and generates the perfectly well-regulated organization. It tells of people, files and (to go beyond Weber) machines which play allotted roles; it tells of hierarchical structures of offices with defined procedures for ordering exchanges and rational division of labour; and it tells of management as the art of planning, implementing, maintaining and policing that structure (p. 77).

Within this mode five schools of thought will be discussed—bureaucracy, classical management, scientific management, decision making and finally, the systems, contingency, culture view of our own era. All five models emphasize structure rather than agency, or organizational function rather than individual action.

Bureaucracy

One intellect that has profoundly influenced thought about the administration mode of ordering was that of Max Weber. Although he wrote at the beginning of this century, his work was not available in English until the 1940s. He therefore had little impact in English speaking countries before the Second World War. Nevertheless, his stature in sociology, alongside Marx and Durkheim, was such that his model of the ideal bureaucracy heavily influenced the later debate about the nature of public management (Clegg, 1990). Weber (1948, p. 214) recognized that organizations were the key institutions of modernity, and that bureaucracy, because of its 'purely technical superiority over other forms of organization' would be its mode of ordering. Weber outlines the essential features of an ideal type of bureaucracy. In essence, each bureaucrat would be guided by the principles of rational legal authority in the form of rules, hierarchy and written records. He or she would work within organizations which would adopt a strict code of conduct governing their employees using the elements listed below (Weber, 1971).

- They should be subject to authority and control in relation to their official obligations only.
- They should form hierarchies with differing authority.
- Each position should be clearly defined.
- The positions should be filled by free selection in accordance with technical qualifications.
- The incumbent should be paid a salary and is assumed to have no other employment.
- Promotion in the incumbent's career should be in accordance with seniority or merit.

The potential power of the bureaucratic organizational form worried Weber. He was concerned that these organizations would become the iron cage of the modern world with procedure dominating substance. His pessimism did not diminish as he grew older. Late in his life he said in a speech

> Already now, rational calculation is manifested at every stage ... this passion for bureaucracy is enough to drive one to despair (Mayer, 1956, p. 127).

He need not have worried unduly, as natural human inefficiency has ensured that the iron cage is less than secure.

In the first half of this century, two other strands of theory dominated thinking about organizations in the English speaking world—classical management theory and scientific management.

Classical management

Theorists such as Fayol (1949) drew on their own experience of well-managed organizations to develop principles of organizing built around a number of key concepts (Morgan, 1986, p. 26)

- *unity of command:* employees should report to one person only
- *schalar chain:* this is the line of authority from the top level in a hierarchy to any one employee
- *span of control:* the number of people reporting to any manager must be limited if good coordination is to result
- *staff and line:* staff advise and line manage
- *initiative:* to be encouraged
- *authority and responsibility:* go together
- *centralization:* the degree of centralization of authority should be varied to achieve maximum efficiency
- *discipline:* rules and norms are to be obeyed
- *subordination of interests:* the organization comes before the individual
- *equity:* fair pay for a fair day's work
- *stability of tenure:* without stability employees cannot develop their skills for the good of the organization
- *esprit de corps:* harmony is the basis of corporate strength.

Many, if not all of these principles can be found in modern organizations, and such is the world that many of them have become so taken for granted that they are now organizational ideologies that support the interests of managers (Alvesson, 1987). The visual expression of the first four principles can be seen in the familiar organizational chart, with its line managers connected to a limited number of staff and staff managers shown off line. In modern organizations the divisional arrangement of large organizations is shown in the same way. Morgan (1986) also points out that mechanistic management techniques such as management by objectives are also the result of this structuralist vision of organizations.

Scientific management

Frederick Taylor was an American engineer contemporaneous with Weber. He advocated five principles (Morgan, 1986)

1. managers should plan and design the work and workers should do it
2. use methods such as time and motion study to find the most efficient way to do a job
3. fit the right person to the job
4. train the worker to do the job
5. monitor performance.

These principles of organizing can be found in organizations as diverse as fast food chains and car manufacturing plants. Tasks are broken up into specialized units of work all interconnected and controlled. Control was (is) the key word. Men and women were expected to become efficient adjuncts to machines and subject to similar surveillance. It is not surprising that Taylor's principles became as popular in the Soviet Union as in Ford's car factories.

Flawed decision making systems

Herbert Simon and his colleagues at Carnegie–Mellon University developed in the 1940s and 1950s a distinctive view of organizations of great relevance to our enterprise (Simon, 1947; March and Simon, 1958; Galbraith, 1977). For them, the human capacity to process only limited amounts of information was reflected in the forms and functions of organizations (Morgan, 1986). Thus, the central decision making function of organizations is simplified by the adoption of hierarchies of decision responsibility, divisionalization to isolate different types of decisions, and rules to constrain and guide the process. Thus local, rather than global goals, tend to dominate the process of organizing. Moreover, organizations respond to increasing external uncertainty by increasing the emphasis on goals and judgement and diminishing reliance on standard procedures and regulation. However, external uncertainty creates greater internal discretion and consequential decision uncertainty. To minimize this undesirable external uncertainty organizations go to great lengths to reduce market competition through agreements or mergers, attempt to control resource flows, and gather intelligence on competitors and markets. Strategic planning may be seen as part of this uncertainty reduction exercise (Cyert and March, 1963). It is worth noting that the image of an organization as a constrained decision making system has echoes in actor-network theory to be described later.

Classical, scientific, and (to some degree) decision making management principles assume that organizations are closed systems with explicit goals. They can be designed using rational principles to maximize output using methods such as division of work, mechanization and standardization. However cold and heartless this conception of organizing is, it was still assumed to be the result of deliberate human design. No such assumption is made in the following theories.

System, contingency and culture

In 1950, the sociology of structural functionalism and general systems theory were combined into a model of organizations which emphasized the interrelatedness of organizational functions and units (Silverman, 1970; Clegg, 1990). The organization was seen as a sort of living thing—an organism with needs, a stabilizing central value system, and the ability to react and adapt to changing environmental forces. Structural functionalism emphasized the nested nature of society and how each subsystem contributed to the functions of the whole—such as, for example, how the family subsystem contributes to the urban system. Central values stabilize systems. In the case of organizations these values are called the corporate goals (Parsons, 1956). General systems theory developed from biology and pointed to the similarity between all systems. All need inputs, all manipulate those resources in some way and all produce outputs. Organizations are systems with boundaries that are free to allow a flow of inputs and outputs to or from the surrounding environment. Feedback from this environment causes the system to adapt for survival.

It is easy to see how this philosophy of the organization developed into a fascination with the relationship between the system and its contingent environment. This body of research, known as contingency theory, explained organizational form in terms of such things as product diversity and company size. Donaldson (1985) has even postulated that organizations should fit one of only thirteen forms (depending on the nature of the organization's role and its market) if it is to achieve maximum efficiency. Contingency theory has continued to dominate much of management thinking about organizations for good reasons (Willmott, 1990). Rapidly changing environments have forced organizations to change internally. Techniques such as strategic planning place great emphasis on an analysis of strengths, weaknesses, opportunities and threats to enable top management to devise ways of structuring the organization to fit the changing needs of the market. The uncertainty of the market, its changing size and content, the new demands of legislation and public sensitivity have caused much introspection in modern organizations. Change now seems normal. The balance of workforce types, the technology to be used, whether to centralize or diversify are all preoccupations generated by a highly contingent environment.

The systems view of organizations can be related to an obsession in the literature with the health of the organism. However, some of the research work on the satisfaction of the human needs of employees started well before the systems view became dominant. The Hawthorne studies of the 1920s and 1930s demonstrated how important the social needs of the workforce was to performance, how friendships created an informal organizational structure and how this could frustrate the plans of classical management (Morgan, 1986). From this period the human resources movement gained prominence as a set of techniques to motivate workers such that productivity increased and harmony prevailed in the organization. Although the link between increased efficiency and increased happiness was often not obvious, the great influence of this point of view has continued because of its natural place in the dominant systems view of organizations. The satisfaction of needs and lately, such procedures as participative management, job enrichment, work teams and in-house training have all been part of an underlying philosophy that happiness and unity of outlook are the keys to organizational health. The organism must have a strong central value system and workforce harmony is the reason for its survival (Morgan, 1986; Bolman and Deal, 1991).

This preoccupation with harmony and productivity has taken another form. The rise of Japanese industry and the great differences between the institutional forms of the individualistic West and the collectivist East has increased our awareness of national culture as a variable in organizations. Japanese national culture is reflected in Japanese organizational culture. The corporate family is all important, individuals must put the collective first and always, and common values must be learnt and internalized (Morgan, 1986). The superiority of modern Japan in many markets produced a flurry of research in the 1970s and 1980s into the production and maintenance of such corporate cultures. However, when translated to the USA and Europe the deep social structures that underlie these practices cannot be found, and what is left is a rather pathetic obsession with corporate mission statements, logos and ritualized company picnics. Conflict and politics are ignored, and the culture most desired by the chief executive becomes the symbolic ideal of management. Of course, organizational cultures do exist but rarely in the form of a unitary harmonizing value system directed towards an agreed set of goals. Multiple cultures are the reality, only one of which is that shared by the top management elite (Bloor and Dawson, 1994). The popular literature of the corporate culture movement reflected Peters and

Waterman's (1982) claim that shared values are the source of a corporate culture. However, one of the few serious empirical studies in this area (Hofstede *et al.*, 1990) found that rather than *value*, it is shared *perceptions of daily practices* that form a culture in organizations. Values were personal and unrelated to the organization's goals. It is the shared practices of organizing, usually put in place by the founders, that distinguishes one organization from another. It is the organization's rules of the game (how we do things) that counts. This is clearly a considerably more superficial degree of socialization than the internalization of national cultural values that takes place in early childhood and is reinforced by the behaviour of a whole national population over a lifetime. It is for this reason that the term modes of ordering is used in this chapter rather than culture.

The enterprise mode of ordering

> Enterprise tells stories about agency which celebrate opportunism, pragmatism and performance. ... So the perfect agent is a mini-entrepreneur. ... fragments of structure add up to a set of opportunities, a set of resources. The stories of enterprise tell that people are driven by self-interest (Law, 1994, pp. 75–6).

This is a very different world from that of organizational systems and of unified corporate cultures. It is the world of agency and action, of interests and conflict, of politics and power. It is about people and technology in dynamic interaction, of hope and fear, and ambition. Because of its emphasis on the individual, much of the detail has been discussed in the previous chapter. However, it is reviewed here to show how it differs from the administration mode.

Conflicts of interest

A good summary of the differences between the administration and enterprise modes of ordering may be found in Morgan (1986).

In his comparison he calls these modes 'unitary' and 'pluralist' but I will take the liberty of using our own familiar labels. How do these two modes view interests, conflict and power (Table 2)?

In organizations where the enterprise mode of ordering dominates, the critical factor at play is the dynamics of interest conflict resolution. However, although tooth and claw are much in evidence, the organization does not disintegrate into full-scale conflict. Somehow the organization survives and is not torn apart, somehow the coalition achieves its ends. Two mechanisms are at work to help the process. Firstly, the administration mode of ordering is working away to control the conflicts using rules and smoothing devices. In particular, it is one of the tasks of management to resolve the inevitable conflict between corporate subgoals. Although all may agree that, for example, maximizing return on investment is the global goal of the organization, each subunit may not have unit goals that contribute directly with this goal or indeed be aligned with it. What is a goal for one unit may be experienced as a constraint by another. For example, retiring debt may frustrate production goals, and rapid expansion to meet sales targets may abort the quality improvement campaign (Simon, 1965).

Table 2. Contrasting modes

	Administration mode	Enterprise mode
Interests	The organization must achieve a set of common goals and pull together to achieve them.	Many individual and group interests exist in an organization. The organization is a loose coalition of these interests.
Conflict	Conflict must be removed and replaced with harmony.	Conflict is a natural and positive part of organizing.
Power	This view talks of authority, leadership and control, and tends to ignore power concepts.	The resolution of conflicts of interests is attributed to power drawn from multiple sources by groups and individuals at all levels.

To some extent the organization may be looked upon as the dynamic end result of 'the resolution, at a given time, of the contending claims for control, subject to the constraint that the structures permit the organization to survive' (Pfeffer, 1978, p. 224). Thus, coalitions of interests and powerful individuals are in an ongoing struggle for power and position, and many of the features that we recognize as structure are largely the temporally stabilized end result. In organizations, the most powerful coalition is usually that of top management, so it is not surprising that organizations largely reflect their notions of order and efficiency. However, anyone familiar with attempts to change organizational structures will be very aware that rationality does not explain the detail. Some divisions may be receiving a share of resources disproportionate to their effect on the productive process and others are irrationally starved. For example, in many production facilities, more attention is paid to the welfare of management groups with little effect on the production goals than to those in charge of engineering functions. This is a reflection of relative power.

Secondly, managers learn to manage personal conflicts of interest using a variety of techniques. These strategies will vary with circumstances and personal style. Thomas (1977) has noted that managers will often use assertive techniques in an emergency or, when an integrative solution is required, suitably collaborative methods are used requiring problem solving methods. Negotiation is resorted to only if compromise is the only solution or more assertive tactics may be damaging. Sometimes, the conflict may be avoided, hidden, or sidestepped if circumstances permit. Finally, some managers deem it wise to admit error and give in—the credit can always be used later. These are not unlike the favourite persuasive strategies of managers in non-conflictual situations, as was shown in chapter 1.

The picture emerges of individuals or groups expressing their particular identities, meanings and ambitions within an administration mode of ordering which is pursuing its organizational role of integrator and controller. These agents are often more interested in their own destinies than the goals of the organization. Nevertheless, they often provide the dynamism of the organization, the raw energy required in an uncertain and ambiguous world. They can be both leaders and destroyers, builders or spoilers and often, both at the same time. The enterprise mode of ordering is as intimately associated with change and decision making as the administration mode is with stability and rules. Both contribute to the flux we call organizations, both are present at all times and both are manifest in all agents' behaviour at different times and in different circumstances. Theories that express this dynamic interaction are discussed next.

The network perspective

Diminishing organizational bureaucracy and the necessity for increased internal and external flexibility has encouraged organizational theorists to model organizations, and relations between organizations, as social networks (Nohria and Eccles, 1992). The analogy of a network serves to preserve the identity and intentionality of actors but, nevertheless, sets them within a web of interactions and relationships. As Nohria (1992) explains, all organizations or other social units may be assumed to be formed from dynamic networks. The nodes of the networks may be individuals, groups or organizations, depending on the level of analysis. The relations between the nodes may be based on friendship, formal relations or may be purely communicative. They may mimic the organizational hierarchy or completely ignore its existence. They are dynamic and changing, as the boundaries move and actors enter and exit. Relations within the network may be singular or multiple, simple or complex, stable or emergent and they interact and interconnect in complex ways in many dimensions. From a macro stand point, organizations are self defining clusters of networks with their environment delineated by the extent of their relations with other organizations or individuals in their organizational field (Di Maggio and Powell, 1983). These fields may include customers, suppliers, competitors and regulators.

Above all, networks constrain and enable action. Actors are embedded in a network of interactions which 'are constantly being socially constructed, reproduced, and altered as the result of the actions of others' (Nohria, 1992, p. 7). Within this network, actors are striving for control and advantage, and as a result are creating new network elements.

> Therefore networks are as much process as they are structure, being continually shaped and reshaped by the actions of actors who are in turn constrained by the structural positions in which they find themselves (p. 7).

Actor-networks

A model of organizing that reflects this dynamism and complexity is the actor-network model which has been largely developed in the last decade by Latour (1987) and Callon (1986) in the School of Mines, Paris and recently utilized in organizational theory by Law (1994). The actor-network concept is developed from empirical studies done in the sociology of technology and science and it is distinguished by its refusal to privilege human actors over technology in the dynamic of organizing. In any particular situation the human and non-human actors form a dynamic interactive network of influences. A key process in this interaction is the ability of a powerful actor node to define the situation in terms favourable to their own interests, enrol the other actors into this point of view and thereby control their reactions to events (Latour, 1987; Callon, 1986; Law, 1992b). These 'centres of translation' or 'centres of ordering' (Law, 1994) control the network, at least temporarily, by channelling, selecting and monitoring the flow of interactive processes, information and technology within the network.

The process of organizing takes place utilizing temporally stabilized actor-networks which include some of the individuals, groups, committees, coffee machines, computer reports and buildings which may be identified as part of the organization, together with the complex plurality of involved actors outside the organization. At any one time many networks exist of different composition and magnitude which inevitably interact to produce effects of which only some will be predictable.

With this understanding we have come a long way from the rationality of classical theorists and the systems view. Here, humans and machines are interacting in complex ways, with the human actors driven by expressions of organizational meaning sometimes in one mode of ordering and sometimes in another. Sometimes the interests of the organization, as interpreted by individuals or groups at particular times, may be dominant and at other times the interests of the individual may be the force behind action.

Implications for decision making

It can be concluded that the administration mode of ordering and the enterprise mode of ordering will require a rather different approach to decision making. The first will tend to mobilize rules and norms as the basis for choice and the second will hope to use problem solving strategies. Other research has indicated that this duality is the dominant pattern in organizational decision making (March, 1994). Law (1994) has indicated that other modes of ordering may exist and naturally these will spawn different decision styles—the 'vocation' mode of professionals will probably try to use linear rational means and the 'vision' mode of charismatic leaders will be defined by the decisions of a single dominant actor. Yet, as we know, any one organizational decision will involve a wide range of people and organizations, with widely different interests and objectives. We can expect, therefore, to find both the administration and the enterprise modes of ordering expressed in the process, and possibly others. Any effective decision methodology must be flexible enough, therefore, to accommodate a number of decision rationalities simultaneously.

Summary

- The study of organizations may be seen as a struggle between forces associated with structure and those resulting from individual agency.
- In line with this point of view, two dominant modes of ordering may be detected. The administration mode of ordering talks of structure, hierarchy, function, control and organizational adaptation, and the enterprise mode speaks of meaning, interests, power, politics and symbolism.
- Theories of organization reflecting the administration mode of ordering include
  - Max Weber's theory of bureaucracy
  - classical management
  - scientific management
  - flawed decision making theories
  - system, contingency and culture theories
- Theories of organizations reflecting the enterprise mode of ordering are less easily classified into schools of thought but may be contrasted with administration mode theories. Thus
  - the unitary, goal driven organization is a fiction. The reality is a loose coalition of individual and group interests
  - conflict is a natural and positive attribute of organizations
  - whereas the administration mode theories tend to ignore power, those associated with the enterprise mode take power seriously, together with its expression in conflict resolution
- Network theories, particularly actor-network theory, attempt to resolve the differences between agency and structure by modelling the organization as a dynamic

network of interactions between individuals, groups and technology.

Although the two dominant modes exist simultaneously in all organizations, we must expect them to spawn different decision making styles. Those operating within the administration mode will tend to rely on rules and norms and those within the enterprise mode will tend to concentrate on problem solving strategies.

3

Engineers and management

Industrialism and positivism

The present relationship between engineering and management cannot be understood without an understanding of the forces that moulded modern sociotechnical world views. This will require us to briefly review the development of professional engineering in the nineteenth and early twentieth centuries and the evolution of its peculiar rationality. But before we do this, it would be prudent to touch on the way modern Western societies as a whole developed their particular attitude to life.

Rapid economic change, brought about by the increasing concentration of capital into the industrialization of agriculture and production, took place in the late eighteenth and early nineteenth centuries. Whole populations were displaced from the land and the cities ballooned in size—particularly where production was now concentrated in the new factories. A new social order began to spread. The commercial life of the villages was being replaced by the fragmented and contingent existence of the city. Life for many became split between work in a factory and the desperation of holding a family together in a violent and disease-ridden slum. Power was concentrated in the cities. Government bureaucracy spread through a proliferation of regulations and the greater ease of communication. Regional diversity diminished and class divisions, based on industrial work, structured social discourse. Throughout the nineteenth century public administration and the new industrial organizations became more rationalized around specialized work roles. This, however, was but a reflection of a new rational world view formed in the eighteenth century and reinforced by an increasing faith in the scientific project (Kumar, 1986, 1988). The cultures of the rapidly industrializing countries changed, but not without resistance. Large areas of Europe and North America were still dominated by agriculture throughout the nineteenth century and were often bypassed by the spreading rail systems. However, the industrial way of life touched more and more communities and the general culture was forced to adjust to the demands of technology. Over successive generations the technology culture, the legacy of the older agricultural, and the new free enterprise capitalist cultures were brought into alignment (Hill, 1988). Now our modern world view is enframed by the cultural text of technology and we cannot think or act effectively outside its context.

Those who drove these changes were inspired by the advances of science and the principles of positivism. This approach to life had at its centre an attitude of instrumental control over nature inspired by a realistic, non-mentalistic view of the world. Facts and values were to be separated, with the former being the only valid basis for rational action. Science was the key to both technical and social progress and technology was its chosen instrument (Kumar, 1986; Fischer, 1990). Moreover, 'social problems conceptualized in technical terms, are freed from the cultural, psychological, and linguistic contexts that constitute the lens of social tradition ... to be technically and administratively engineered by experts' (Fischer, 1990; pp. 42, 44). Many of these experts were engineers.

The engineers

Reports of the development of industrial works in the late eighteenth and early nineteenth centuries are replete with references to engineers (Berg, 1994). However, these were not the gentlemen designers of later years but rather the inheritors of the skills of the millwright. These artisans worked in iron rather than wood in the new occupation of machine building and maintenance. Yet this body of practical knowledge of materials and mechanisms became the source of much later design work. Similarly, the canals, bridges, tunnels and railways of the eighteenth and nineteenth centuries were built by men of no formal engineering science education but, nevertheless, possessing great practical skills obtained through pupillage and trial and error. Inevitably, those with the education, money and entrepreneurial skills to actively cooperate with the new capitalists in the design and production of civil and mechanical artefacts wished to differentiate themselves from mere artisans. Thus, at the 1818 founding of The Institution of Civil Engineers, (ICE), in London, it was proposed that

> An Engineer is a mediator between the Philosopher and the working Mechanic, and like an interpreter between two foreigners, he must understand the language of both (Lloyd, 1991; p. 31).

Sensibly, this definition both flattered the *philosophers* of the new scientific world, firmly established an intellectual dominance over the artisans, and carved out a position independent of both science and craft. Yet, in reality, the intellectual edge over the craftsmen was small. In the English speaking world, it was not until the latter half of the nineteenth century that engineers required other than practical training to claim professional status, and by that time the *mechanics* were often widely read (Thompson, 1980). The claim to professional status was boosted when, through the influence of Thomas Telford in 1828, the ICE was awarded a Royal Charter. Thus, British engineers entered the strange world of professionalism, to be followed some decades later by those in America.

As the twentieth century progressed, design in particular became increasingly codified. Scientific research into materials and mechanisms had produced useful data and this, together with the mass of rules of thumb accumulated over a century and a half of practice, produced a proliferation of standards, codes and specifications to govern the processes of engineering. These rules of conduct have been updated over the years to reflect the increasing scientific data and the new demands of industry, but nevertheless, much of routine engineering design remains not so much an application of science as a compliance with norms of good practice (Brown, 1984).

The production and construction processes have, of course, been strongly moulded by standards and specifications. However, much of the practical day-to-day management of production and construction has remained faithful to the trade origins of engineering. The direct control of production and construction in the English speaking world is still largely performed by managers with technical roots but without engineering degrees (Glover and Kelly, 1987; Lloyd, 1991). It is only as the management function moves further away from questions of practical engineering and the direction of workers, are we likely to see more tertiary qualified engineers involved in management tasks. The perception among engineers is, however, that the ability of professional engineers to penetrate upper management levels has been blocked. This is a question to which we will return later. In the meantime, we will address the matter of professionalism.

Professionalism

The habit of forming occupationally-based, government-recognized associations seems to have a particular attraction in countries with past cultural ties with Britain. Knowledge-based occupations in much of Europe and Japan seem to exist without these private institutional structures (Collins, 1990a). Therefore, we will discuss professions as experienced in the English speaking world—particularly in Britain and America.

The key to the formation of a profession is not the possession of a body of abstract knowledge. Professions such as civil engineering were established well before science dominated the knowledge base. The answer lies in the ability of group associations to assert 'hegemonic control' (Hill, 1988) because of state recognition of certain occupational privileges. Today, of course, protection of these privileges is achieved by an examination system regulated in some way by the professional associations. Thus market closure is achieved (Collins, 1990b). This closure is further reinforced by the careful elaboration of status using rituals and codes of conduct. For example, the ideals of service and autonomy are used to differentiate professionals from both industrial capitalists and trade union groups, and help, therefore, to continuously re-emphasize its contentment with its non-threatening position of privilege. Thus,

> The new professional men brought one scale of values—the gentlemen's—to bear upon the other—the tradesman's—and produced a specialized variety of business morality which came to be known as 'professional ethics' (Reader, 1966; p. 158).

Professional privilege must be defended assiduously. In particular, the control over the knowledge base is continuously being challenged by other occupations—nurses demand more say in treatment, and structural design is being penetrated by technicians. Jamous and Peloille (1970) considered that the long-term invulnerability of these bodies of knowledge depended not so much on its complexity as on the degree to which judgement is required in its application. For example, the more rule-bound design becomes the more likely it is that design will be performed by para-professionals. Medicine overcomes this difficulty by spawning greater and greater interactive layers of detail through scientific research, much of which is introduced quickly to the knowledge base of specialists and eventually to that of the general practitioners. Similarly, rapid changes in legislation keep lawyers and accountants one step ahead of the general public.

In professional engineering, ideals common to other professions may be found. A respect for the body of knowledge, a feeling of loyalty to fellow engineers and a pride in the craftsmanship of the calling are all echoed in the views of other professions. Although engineers have rarely enjoyed autonomy of practice, they still have the privilege of deciding how to do an engineering task.

Unfortunately, for most engineers, the pride in the status supported by these ideals has rarely been translated into organizational power. Why is this?

Some clues may be found in the different demands of design engineering and engineering management. The role of the modern design engineer is supported by a combination of practice-based rules and the body of knowledge derived from engineering science. These have been embodied in the standards and codes of practice which guide the creation of design documents and specifications. These, in turn, have been underpinned by the active pragmatic and materialistic philosophy of positivism. Indeed, design engineers may be looked upon as ideal positivists, content to fashion brute matter in whatever way desired by their industrial masters. Positivism gives the design engineer that untroubled certainty that 'their mission in the world is to introduce the order which they see as facilitating the modernist culture' (Brown, 1984; p. 38). The design engineers' quiet detachment is also supported by a feeling that the performance of their duties, in accordance with their professional code of conduct and its physical embodiment in the standards and codes of practice (partly designed to protect the public against unnecessary risk), is sufficient to legitimize their role. Engineering designers take pride in being the problem solvers—it is not for them to question the nature of those problems.

For engineering managers, however, the conversion of raw materials into useful objects is a messy, often brutal exercise in conflict, persuasion and power. Human beings and their interests are at the heart of this project. Not for them exists the cool detachment of the designer; for them positivism is not a useful philosophy. Human emotion cannot be ignored—human beings are not inert objects, and unlike billiard balls, they can decide their own direction. Indeed, the selection for positivism inherent in the school and university system may fatally handicap an engineering manager. He or she will waste time and resources looking for the one best way, or making judgements on the basis of irrelevant but hard data such that judgement will be stunted by the exclusion of soft (human) data. Finally, the perversity of humanity may drive the young positivist back from human interaction to the cool safety of the computer.

For the engineering manager the ideals of professionalism may also be something of a handicap, for they will clash with the ideologies of management. Alvesson (1987) considers much of the organizational and managerial world-view to be ideological. By an ideology he means a shared framework used to legitimate group behaviour, and to serve the interests of that group (Hartley, 1983; Craib, 1992). Because ideologies serve powerful interests they have an element of strategic distortion associated with them—not necessarily false beliefs but certainly partial.

Alvesson (1987) has discussed a number of management ideologies which may potentially conflict with professional views. For example, the ideology that all members of an organization have the same goals and values, and that conflict is local and transient, is a fixation of many upper management groups. Of course, this is a distortion of reality, and a distortion that may put professionals in a poor light. The exemplary professional will look to the profession rather than to organizational goals for a sense of work meaning; work methods will be derived from a knowledge-base inaccessible to many top managers; and he or she may worry about the interests of the general public. These are not the concerns of an ambitious manager for whom the organization provides all necessary sources of work meaning. Similarly, the managerial ideology that considers managers as an elite group with competencies that naturally set them above all others, is unlikely to sit well with professionals. They may perceive that often managers are poorly educated, chronically incompetent and ridiculously overpaid. As Raelin (1985) has noted, professionals see managers as restricting them to specialist roles, supervising them too closely, requiring of them an unnecessary respect for the hierarchy and its goals, imposing conformity through teamwork, and devaluing quality in the name of efficiency. Naturally for managers, these professionals are an unrespectful problem group that stands in the way of corporate harmony and efficiency.

But these ideologies are not the only source of friction. Managers are often nervous about the possible encroachment of well-educated professionals into their domain. This threat may suggest that much of the management talk about difficult professionals may be generated by their own conscious or unconscious market closure mechanisms (Larson, 1977). Professionalism may be seized upon as a very good reason to divert engineers and others on to a dual career ladder, well away from line management. Not only is the problem of professionalism used to legitimate this action, but the very nature of the professional training may be cited as a further good reason to restrict professionals to their specialized roles. For example, Armstrong (1987) has quoted the tendency of top British managers to classify engineers as unsuitable for management because of the narrow nature of their education. But why then, have some professions done better than others in their efforts to break out of the specialist box and into the ranks of top management? For clues we should turn to the work of Armstrong (1984, 1987) who has examined the reasons for the failure of British engineers to penetrate senior management ranks.

Armstrong considers the core of the problem to be in the concept of management which dominates English speaking organizations. This places the real managers in functions remote from the details of production and in a role concerned with the profitability and market share of business rather than the nature and manufacture of the product. This disdain for nuts and bolts has been reinforced by an education system which has produced 'a whole generation of MBA students who will not go near a manufacturing strategy—they want to be in at the gin-and-tonic end with the financial strategy' (Parnaby, 1985; pp. 43–5). In this heady atmosphere engineers are not considered suitable management material. In contrast, German and Japanese engineers can be found at all levels of equivalent organizations. Armstrong attributes the difference to the very different resource-base in Germany and Japan. From the earliest of modern times, in Britain and America the securities market produced much of the capital required by industry and as a consequence the control and audit of money flows became an important function of management. Therefore, accountants became central to the welfare of English speaking capitalism. In contrast, finance in Germany and Japan was largely obtained from banks and the government, leaving considerably more organizational energy for the management of production. One of the happy consequences of this was that relatively few accountants (and lawyers) were to be found in large Japanese and German companies. In Germany, in particular, where management was dominated by engineers, simple but effective techniques of financial accounting were assumed to be part of the knowledge-base of all professional engineers. However, we can assume that the modern internationalization of markets and production will steadily increase the influence and number of accountants in all countries.

In the English speaking world, the control of money flows and the complexity of public company audit and taxation law ensured the accountant's position in the ranks of senior managers. The growth of large corporations in America and elsewhere consolidated this position. The control of the allocation of resources within these companies became a function of financial rather than technical expertise. Money was the common denominator and engineers did not understand the complexities of its manipulation. As a consequence, claims by engineers that all industrial management contained an 'engineering dimension' (Glover and Kelly, 1987) and its neglect has, and would in the future, be detrimental to national welfare, did not unduly influence non-engineers in top management. Indeed, it is clear that to draw attention to the importance of the product may improve the rewards of design engineering, but it is unlikely to recommend engineering as a source of strategic managers.

Competition between professionals for management positions is important in understanding the relative weakness of engineers in the top echelons of organizations. However, it should not be exaggerated. In 1980 in Britain, only between one half and two thirds of managers had a higher education (Armstrong, 1984). It is probably true that a high proportion of managers in other English speaking countries are similarly qualified. What they do have, however, is a firm grasp of those skills required to manipulate people. Management, defined as getting things done through other people, has always been the key skill of organizing and no professional education can compensate for its absence. As we have noted before in this chapter, the positivistic world view of engineers does not predispose them to human understanding. Nevertheless, for those with leadership skills there is no reason to think that a professional education will be a disadvantage in the competition for management positions with those without a profession. And certainly a dry-as-bones accounting education is unlikely to develop better people-management skills than that of an engineer.

Merging roles

It is possible that too much is made of the potential conflicts between the professional world view and that of managers. Certainly engineers have always been embedded in organizations and have learnt to adapt to the appropriate corporate rules of the game. And many have become successful managers. Nevertheless, in the English speaking world it appears that the structures of control in large organizations have not evolved to readily welcome engineers into positions of power. But these control mechanisms are changing in many types of institution, and the consequences for engineering managers are profound. For an analysis of this new situation we will follow that of Burris (1989).

Up until about 1960, Burris considers that three forms of control could be found in large organizations. In factories, a *technical control* dominated which worked through the design of automated, pace-setting production systems. In non-industrial organizations *bureaucratic control*, based on divisionalization, specialization and rationalization dominated. In settings with less routine and higher knowledge requirements a *professional control* system was common, organized around status groups, self-regulation and high levels of training in esoteric skills. Naturally, a mixture of these modes of ordering (Law, 1994) can be found in most organizations. We have already discussed some of the problems that can arise when such mixtures occur.

Burris (1989) considers that since 1960, in technology and data-dependent institutions such as those in banking and telecommunications, all three methods of organizational control have fundamentally changed. They are all merging into a form of modern control and organization she calls *technocracy*. The changes are summarized in Table 3.

Clearly, the most obvious result of computer controlled factory systems has been the shedding of a large number of the workforce. Expertise in the design and maintenance of manufacturing equipment has displaced machine-user skills. Any work that stubbornly resists automation can easily be exported to the third world. Computers now enable production performance to be monitored at any level in the organization and designs can be worked upon simultaneously by experts located in dispersed locations. Re-skilling has become imperative for staff caught up in the maelstrom of technological change. Many tasks can be decentralized provided the work group can access the computer network—the same network that, paradoxically, enables centralized decision making based on rapid data processing.

Reorganization around computer technology has produced equally dramatic changes in large non-engineering organizations, or in the head offices of engineering businesses. Thus, expert-dependent technical systems have replaced middle management to a high degree and hierarchies have flattened. The division between the experts and non-experts have widened and credentialism has become important for promotion. Formal job ladders are disappearing to be replaced by project-based work where competence is the criterium for reward rather than seniority. Technical expertise is now a requirement for management in the new technocratic organization. The management and technical roles are merging. This convergence of professional and administrative functions is muting the old contradictions. However, the rationalization of professional work resulting from advanced computer software has

Table 3. Transformations to technocracy (Burris, 1989)

Technical control	Technocracy
Polarization into (small) conceptual workforce/(larger) non-conceptual workforce	Polarization into (larger) expert sector/non-expert sector
General alienation of workers	Polarization of working conditions
De-skilling	Skill restructuring
Centralization	Flexible configurations of centralization/decentralization
Bureaucracy	**Technocracy**
Hierarchical division of labour	Flattening of hierarchy Polarization into experts and non experts
Organizational complexity	Technological complexity
Internal labour markets; seniority as basis for promotion	External labour markets; credential barriers
Task specification at all levels	Flexible, task-force orientation at expert levels; routinization at non expert levels (typically)
Rank authority	Growing emphasis on expertise as basis of authority
Professionalism	**Technocracy**
Relatively autonomous work situation; colleague control	System integration; administrative responsibility/control
Collegial control of training and gatekeeping	Institutionalized credentialism
Client orientation; ethic of service	Organizational/systems orientation; ethos of efficiency
Broad mystique of competence	Specialization of skill; technical expertise as basis of legitimation

blurred the divisions between professional and para-professional expertise, and left the professional sphere vulnerable from below. Luckily for some, the indetermination of management has enabled them to reassert their judgemental claims of professional status. The management function has now become an essential legitimator for the continuing professional role of engineers and other technical professionals.

Implications for decision making

Engineers are natural positivists and this will lead them to seek neat, unfuzzy solutions to problems. They will hope for models and systems to guide them towards a singular right answer. Human beings, in all their perversity, will frustrate them constantly. Their training is in solving mechanical problems using well-established rules, not in diagnosing complex problematic situations and managing solutions using guile and politics. So, management decision making may be a stressful and puzzling experience for an engineer. Their action-oriented outlook will not cope well when solutions cannot be found or are ambiguous. Their professional background will not help and indeed may bring them into conflict with other managers. Bitterness at exclusion from the top decision making ranks may cloud their judgement. Yet opportunities for real decision making roles may well be increasing for the engineer. Technology is changing fast and the old bureaucratic order is crumbling. Expertise, not rank, is the key to success in the new technocratic organization. Increasingly, it will be the young that make the crucial sociotechnical decisions in the new project-based organizations. It is best to be prepared.

Summary

- The modern technocratic view of life is a natural product of the progress of industrialization, and engineers share a particular codified form of this world view which is encouraged by the professionalization of the work.
- The positivistic and codified training of engineers offers little guidance to effective behaviour in the messy human world of management. This disadvantage is exacerbated

by the historical exclusion of engineers from top management positions in English speaking countries.
- Provided engineers have the capacity to overcome these problems, new organizational forms are providing increased opportunities for them to become involved in organizational decision making at all levels.

Part II

Analytical decision making

Most engineers are happiest when problems can be modelled in some way. Accordingly, decision theory in its mathematical form is introduced in this section of the book which discusses some analytical decision aids that may be useful to engineering managers. Chapter 4 examines the theory and practice of decision analysis, a technique commonly found in the business world which explicitly acknowledges uncertainty. Decision analysis is normative rather than descriptive—it models how numerically based decisions should be made if we are to fulfil the axioms of probabilistic rationality. Because it uses probability theory it is particularly useful for decisions where forecasting is important. It is admitted, however, that decision analysis does not mimic human decision strategies very closely and care must be taken in its application.

In contrast, chapter 5 discusses the pros and cons of choice techniques and rules that do not require consideration of probabilities. Important techniques of analysis are discussed—techniques that require complex potential solutions to be disaggregated, assigned weights and scales, and through various numerical methods, assigned a cardinal ranking. These techniques are particularly useful where choice requires the balancing of a large number of potentially conflicting problem attributes, and they have found a useful niche in resource allocation problems and the ranking of projects. Other less numerical techniques are also discussed which utilize common sense choice rules that do not require probability estimates.

Finally, chapter 6 brings probability back into the picture when discussing the management of physical risk. This is a particularly important subject for engineers and is based on the same set of probabilistic assumptions that inform decision analysis. This technical model of risk is very different, however, to the way most members of the public view risk. This sharp division of perception has been one important reason why analytical risk analysis has done little to advance the social management of risk. This chapter is an excellent illustration of the dangers of an overly technocratic approach to engineering decision making.

4

Decision analysis

The ideal of rationality in human thought and behaviour is with us always. Our self image demands that we think of ourselves as creatures responding in reasoned ways to life using the higher functions of the brain. This in fact, is not an unreasonable assumption as much behaviour can be explained using such assumptions. With this ideal image of human functioning in mind we could model the human decision process in the following way

- identify the problem or opportunity
- think about the problem/opportunity with your goals in mind. This thinking is a rational search for data and a reasoned inference based on that data (Baron, 1988)
- at the end of the inference process make a judgement about the situation
- having made a judgement as to the best cause to follow, decide. Decision, in this case, may be defined as an intention to take action
- act.

It should be acknowledged that the key to this decision process is the highly opaque inference/judgement stage, and much of decision theory is an attempt to describe or aid its operations. In this chapter certain modelling aids to the inference process will be discussed. We will assume that the data required for a decision has been gathered and that we are capable of translating our judgement into decision and action. The key to our further progress is a philosophy of rational inference called expected utility theory—a model of human reasoning that has dominated modern decision theory and flowed on to structure thinking in many important areas of sociopolitical enquiry. This expected utility model is an ideal based on certain assumptions concerning human rationality. It is not derived from observation but from theories of rational economic behaviour developed half a century ago (Baron, 1988). It is a normative rather than descriptive theory—it tells us how, as rational human beings, we should make judgements if we are to fulfil certain requirements of economic and probabilistic theory.

Risky choice

When you are deciding whether to spend a few dollars or pounds on a lottery ticket, you will tend to compare the investment with the potential return. If you know the chances of winning are very low you will only invest if the pay-off is very high. On the other hand, if the pay-off is modest you will only invest if the chance of winning is reasonably high. In this process of risky choice you have lived the central principles of decision theory or expected utility theory.

We should start with a few explanations. Utility is the subjective value or desirability of an outcome. Thus, the utility of winning a great deal of money is high because we feel that by doing so we will achieve many of our life goals. Obviously, the disutility of losing the money invested is small otherwise we would not have parted with it. Our probability estimate is also to a large degree subjective. It is unlikely that we know or care about the past frequency of winning a lottery. Certainly, most of our probability statements such as likely or unlikely are guesses based on very limited data. Even probability estimates in engineering are only partly based on a knowledge of frequencies. For this reason expected utility is often expressed as subjective expected utility.

The common sense behaviour of human beings faced with gambles has been used to develop some useful techniques to aid complex decisions. Decision analysis, a technique used in some business decisions, will be discussed below and techniques suitable for more social problems will be looked at in the following two chapters.

Expected utility theory (EU)

EU is a normative theory of the inference process of decision making (Baron, 1988; von Winterfeldt and Edwards, 1986; Watson and Buede, 1987). It is assumed that all the data necessary for the decision is available and we can explicitly define the problem and the goals to be achieved. Of course this sort of assumption is common to much of engineering analysis and, therefore, EU is a very congenial theory for engineers. Expected utility theory is founded on axioms of which the following four are the most important (Wright, 1984)

1. *decidability*—this states that you must be able to decide between outcomes.
2. *transitivity*—if you prefer A to B and also B to C then this axiom states that you must prefer A over C.
3. *dominance*—if all but one outcome of events A and B are equal and just one outcome of A is preferred to B then you must prefer A over B.
4. *sure-thing principle*—when making a choice, outcomes that are not related to that choice, or are common, must not influence that choice.

These principles in combination with other less important axioms lead to a conclusion that the rational optimal choice is the one that maximizes the expected utility of an action. To calculate an EU it is necessary to multiply probabilities expressed as numbers between 0 and 1 and utilities expressed on a scale between (say) 0 and 100. Thus, the EU of a decision calculated as $0{\cdot}001 \times 500$ is the same as the EU of another decision calculated as $0{\cdot}5 \times 1$.

EU principles can best be illustrated using a *decision tree* which plots the possible actions, outcomes and utilities of a decision through time. As an example, let us take a simple decision whether to go to work by bus or car. Thinking carefully we decide that driving in uncongested conditions is the best outcome and we give it a utility of 100. Congestion is, however, very aggravating and could not be worth more than 40 on the same utility scale. Going to work on a bus with a seat is not as much fun as uncongested driving therefore we may assign 80 to this outcome. Alternatively, standing is the worst outcome and gets a mere 30 in the scale. Past experience leads us to guess that the probability of these outcomes on a day like today are 0·7 for congestion, and therefore 0·3 for no congestion and 0·6 for getting a seat and therefore 0·4 for standing. This subjective data can now be organized into a decision tree (Fig. 1).

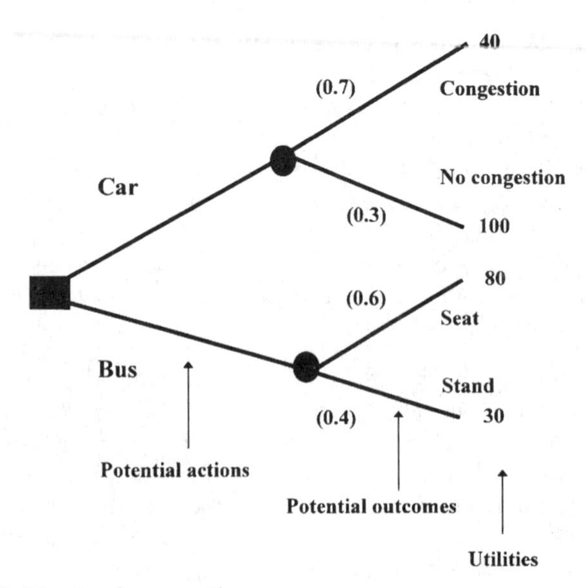

Fig. 1. Decision tree for commuting

We can now calculate the EUs of each potential action.

$$EU\ (car) = 0{\cdot}7 \times 40 + 0{\cdot}3 \times 100$$
$$= 58$$
$$EU\ (bus) = 0{\cdot}6 \times 80 + 0{\cdot}4 \times 30$$
$$= 60$$

Thus, in accordance with EU theory, we should choose to ride the bus today.

Decision analysis (DA)

DA utilizes the decision tree and the probabilistic basis of EU theory directly. Generally, DA uses money as an expression of utility. This assumes, of course, that in a business setting, money has equal utility for all and that its utility is linear. In other words, one thousand dollars or pounds has the same value to all interested parties and two thousand dollars or pounds has twice that value. Hence expected utility becomes expected monetary value (EMV) and is expressed in terms of a currency.

DA can best be explained by using an example. A small mineral mining company has obtained the rights to mine an area where surface samples and substrata maps indicate worthwhile ore deposits. The geologist estimates that there is a 40% probability that a modest ore yield will be found and a 20% probability that a large ore yield will result. A 40% probability is estimated for no yield. The accountant estimates that no yield will result in a loss of \$150 000 at the end of the first year of mining, a modest yield will generate a \$100 000 profit and a large yield \$500 000 profit. The managing director has decided to use DA to help her make a decision and draws the decision tree shown in Fig. 2.

The do not mine option has an EMV = \$0. The EMV of the mine option is calculated as

$$(0{\cdot}4)(-150\ 000) + (0{\cdot}4)(100\ 000) + (0{\cdot}2)(500\ 000) = \$80\ 000$$

As this figure is positive, it is rational (according to EU theory) to proceed with mining. However, the possibility of losing \$150 000 is making her nervous and she is thinking about getting a further pit sample. However, such a sample could cost up to \$20 000 to get and test. Moreover, in the MD's experience, pit samples are sometimes misleading. Her estimate of the reliabilities are shown in Table 4. For example, there is a 20% chance that a pit sample would indicate a deposit when none exists and a 10% chance of a zero result in a good deposit.

How would this expensive test affect her decision tree? Firstly, of course, the pay-offs would be reduced by \$20 000. Secondly, the probabilities would change in accordance with Bayes Theorem. This is a theorem in probability theory that allows the adjustments of prior probabilities on the basis of imperfect tests—and all real tests are imperfect. In its mathematical form it can be expressed in the following way (French, 1989; p. 39).

$$P(sj|D) = \frac{P(D|sj).P(sj)}{P(D|s_1).P(s_1) + P(D|s_2).P(s_2) + \cdots + P(D|s_n).P(s_n)}$$

where D is a piece of data (like a test result) and $P(s_1)$, $P(s_2)$, $\ldots P(s_n)$ are the prior probabilities of certain states of

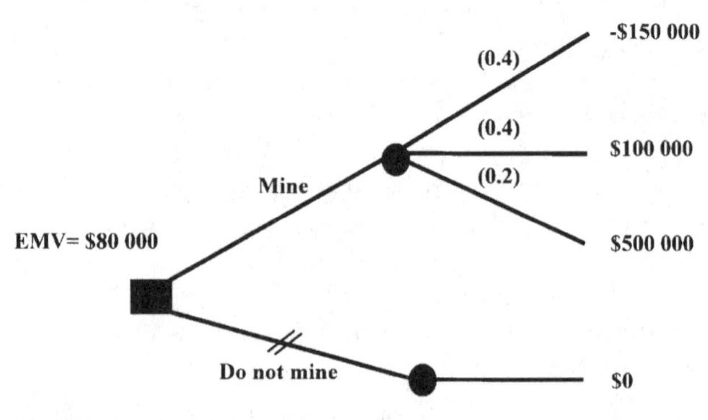

Fig. 2. Decision tree for mining

Table 4. Test reliabilities

Actual	Ore sample results	
	Positive	Negative
No ore yield	0·20	0·80
Modest ore yield	0·70	0·30
Large ore yield	0·90	0·10

nature before D is known. $P(D|s_j)$ is the conditional probability for the occurrence of D given that s_j is the true state, and $P(s_j|D)$ is the posterior probability of s_j given D.

This formula is best understood using a block diagram (Fig. 3) for two prior probabilities.

The revised (posterior) probability of the prior probability $P(s_1)$, is equal to

$$\frac{P(D|s_1).P(s_1)}{P(D|s_1).P(s_1) + P(D|s_2).P(s_2)}$$

In our example the prior probabilities are 0·4, 0·4 and 0·2 which will be modified by the flawed test shown in Table 4. The resulting block diagram is shown in Fig. 4.

Hence, if the pit sample test showed no ore and the test had an 80% probability of being correct, the probability of no ore (0·4) should be revised upwards to

$$\frac{(0{\cdot}4)(0{\cdot}8)}{(0{\cdot}4)(0{\cdot}8) + (0{\cdot}4)(0{\cdot}3) + (0{\cdot}2)(0{\cdot}1)} = 0{\cdot}70$$

Similarly, the prior probability of a modest yield (0·4) should be revised downwards to

$$\frac{(0{\cdot}4)(0{\cdot}3)}{(0{\cdot}4)(0{\cdot}3) + (0{\cdot}4)(0{\cdot}8) + (0{\cdot}2)(0{\cdot}1)}$$

Consequently the posterior probability of a large ore yield drops from 0·2 to 0·04.

Also, if the pit sample test showed the presence of ore the prior probability of no yield (0·4) would have to be revised down to

$$\frac{(0{\cdot}4)(0{\cdot}2)}{(0{\cdot}4)(0{\cdot}2) + (0{\cdot}4)(0{\cdot}7) + (0{\cdot}2)(0{\cdot}9)} = 0{\cdot}15$$

Similarly, the posterior probability of a modest yield would be an upward revision from (0·4) to

$$\frac{(0{\cdot}4)(0{\cdot}7)}{(0{\cdot}4)(0{\cdot}7) + (0{\cdot}4)(0{\cdot}2) + (0{\cdot}2)(0{\cdot}9)} = 0{\cdot}52$$

and the revised probability of a large yield would be 0·33.

The block diagram also indicates that the overall probability of getting a positive test result from the pit is 0·54.

The revised decision tree is shown in Fig. 5. The result indicates that the EMV with the test pit exceeds the EMV without a pit. So

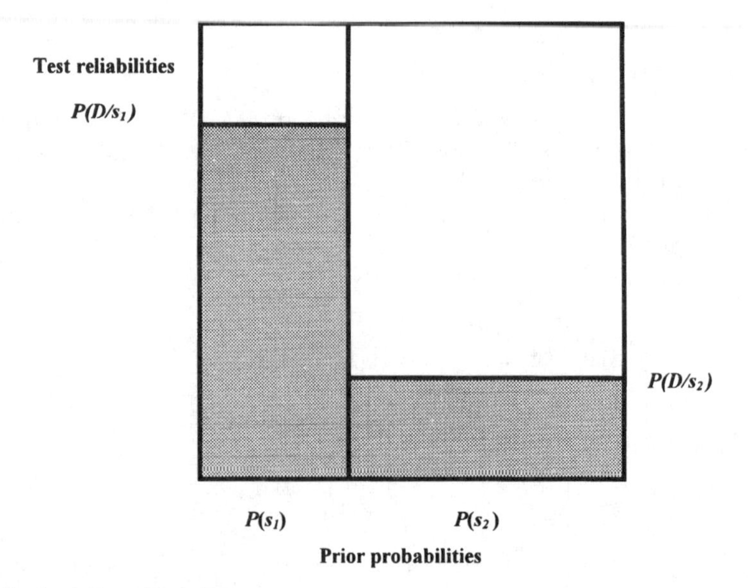

Fig. 3. A Bayes block diagram

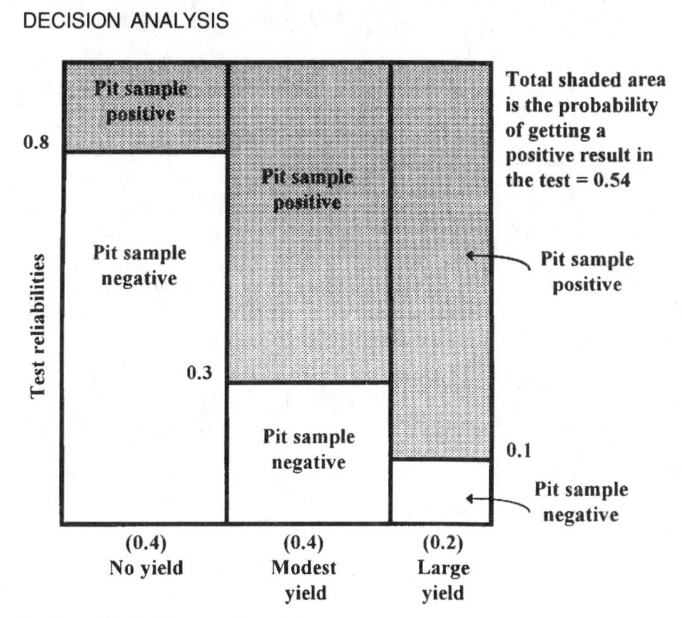

Fig. 4. Bayes block diagram for mining

rationally we should test. It should be noted, however, that should the estimated cost of a test pit have been \$25 000 then the results would not have been worth the expenditure. In the language of DA, \$25 000 would have been the expected value of new information. A further tip—do not accept such a result without investigating the sensitivity of the result to reasonable variation of the pay-offs and probabilities.

For further details on the techniques of decision analysis, you should consult one of the numerous books on this subject, including French (1989), Goodwin and Wright (1991), Moore and Thomas (1976), and Watson and Buede (1987).

How realistic is decision analysis?

The Nobel prize winner, Herbert Simon (1992) describes expected utility theory as 'a major intellectual achievement of the first half of

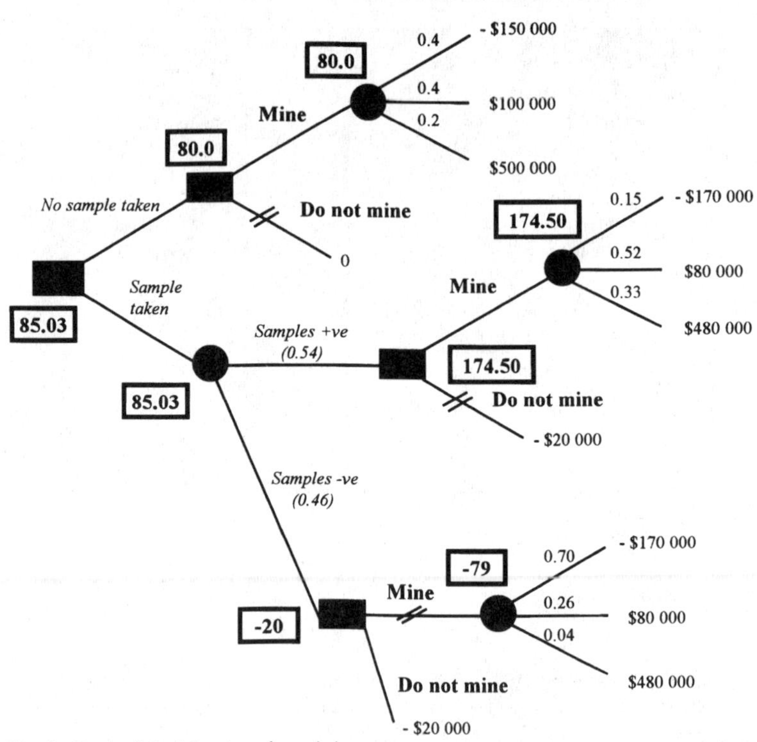

Fig. 5. Revised decision tree for mining

the century' which 'opened the way to fusing subjective opinions with objective data' (pp. 34, 35). He goes on to describe the contribution expected utility theory has made to operations research, including decision analysis, but warns us that its assumptions 'cannot be satisfied even remotely for most complex situations in the real world' (p. 35). Only in very closely constrained problematic situations can methods such as decision analysis be useful—and only rarely will such a situation present itself to the average engineering manager. But, even in these micro-situations, how can we expect to perform as human operators of the decision analysis tool? Not too well, I am afraid. Part of the problem lies with our lack of skill at probabilistic thinking.

Constraints to probabilistic thinking

The list of our failures to conform to the rationality of probability theory is long, but worth reviewing as an antidote to the sin of arrogance. We should bear in mind, however, that despite a certain tendency to overconfidence, human judgement works well in most real world situations (Goodwin and Wright, 1991)—a fact we have explained and exploited in Part III of this book. However, when performing decision analysis we are not acting naturally and the results of the laboratory tests noted in the following section may well be relevant. We can start by describing some of the difficulties people have with probabilities. I will rely on the surveys presented by Mullen and Roth (1991) and Hogarth (1987).

- *Gambler's fallacy.* People tend to believe that independent events are in fact dependent. So, when observing a run of black in roulette many people will bet on the red 'because it has to come up soon'.
- *Availability fallacy.* We tend to overestimate the probability of events of a type that can be easily recalled but underestimate those that are unfamiliar.
- *Scenario thinking.* Instead of looking for objective evidence (e.g., frequencies) we often base our probability estimates on an attempt to imagine what will happen. This may lead us to fall into the availability trap or a form of self deception.
- *Ignoring base rates.* If, when we estimate the probability of an event compared to another, we ignore the difference in population size we can go badly wrong. In an example, subjects were told that 70% of engineers, and 30% of lawyers liked to do mathematical puzzles. Out of a group of 30 engineers and 70 lawyers one person was selected and she was a puzzle freak. Is she more likely to be a

lawyer or engineer? Most people pick engineer on the basis of the puzzle solving and ignore the fact that there are far more lawyers than engineers in the population. If you work it out you will see there is an equal probability of the person being of either profession.

- *Fallacy of small samples.* Because of human laziness, we often base estimates or judgements on quite tiny samples—often only one typical example. This can lead to gross errors when generalized.
- *Fallacy of the law of small numbers.* This is the reverse of the previous fallacy. Here, a person attempts to predict the outcome of a small number of events based on what is known about a large population. For example, if we toss a coin and get the sequence HHHH and later HTTH and are asked which is the more probable we tend to pick the second despite their equal probability.
- *Frequency.* We tend to ignore the unseen failures and only count the successes. For example, a consultant may compare the job lists of two divisions and deem the one with the most jobs to be the most successful at marketing. This ignores the possibility that the 'successful' division may have had to waste vast resources on a long list of failures.
- *Illusory correlation.* We sometimes believe that two events covary when they do not.
- *Conservatism.* If we receive new information, after making a probability estimate, we may not adjust the figure as much as Bayes theorem would predict.
- *Joint probability error.* We tend to grossly underestimate the joint probabilities of several events.
- *Best guess.* Uncertainty may be ignored altogether and estimates made on the basis of a likely value. This ignores the inherent variability of natural events.

No doubt there are other demonstrations of the fallibility of humanity when faced with probabilistic tasks. Indeed, Brehmer

(1980) has suggested that probabilistic thinking is not easily learnt, and consequently we tend to use deterministic, rather than probabilistic inference rules. Causality is apparently developed by our early teens but the notion of chance may never be other than superficial. Our normal experiences do not reinforce the abstractions of probability. Moreover, in pre-industrial groups, it may not develop at all (Wright and Phillips, 1980).

Prospect theory

Apart from our difficulties with probabilistic thinking, it has not been hard for psychological researchers to demonstrate that EU theory is not a good description of human inference (Plous, 1993). Moreover, abundant evidence from the observation of real decision behaviour demonstrates the essentially normative nature of EU theory. (But we must remember that EU theory has, generally, not claimed to be a description of real behaviour.) Some of the problems with EU theory are illustrated in Kahneman and Tversky's (1979) prospect theory article which attempts to modify EU theory based on their extensive collaborative research on human judgement, and the framing effect in particular.

Like EU theory, prospect theory uses two components. In place of utility functions (which in the case of DA are expressed as net worth and are linear) it substitutes a value function in terms of gains and losses. A possible value function is shown in Fig. 6.

In line with the framing effect, people are assumed to judge outcomes in relation to a reference point, which is often the *status quo*. Also, the negative value of a loss of a certain amount is greater than a gain of the same amount, and the value falls off faster. Both sides of the function indicate that our utility is nonlinear and subject to the law of diminishing returns. The shape also indicates that, if an outcome is framed as a gain, the decision maker will tend to be risk averse, and if framed as a loss, will tend to risk taking. Thus, those being sued are likely to be less inclined to settle out of court than those who are looking forward to the gains of litigation. Similarly, the disadvantages of the resignation of your present boss

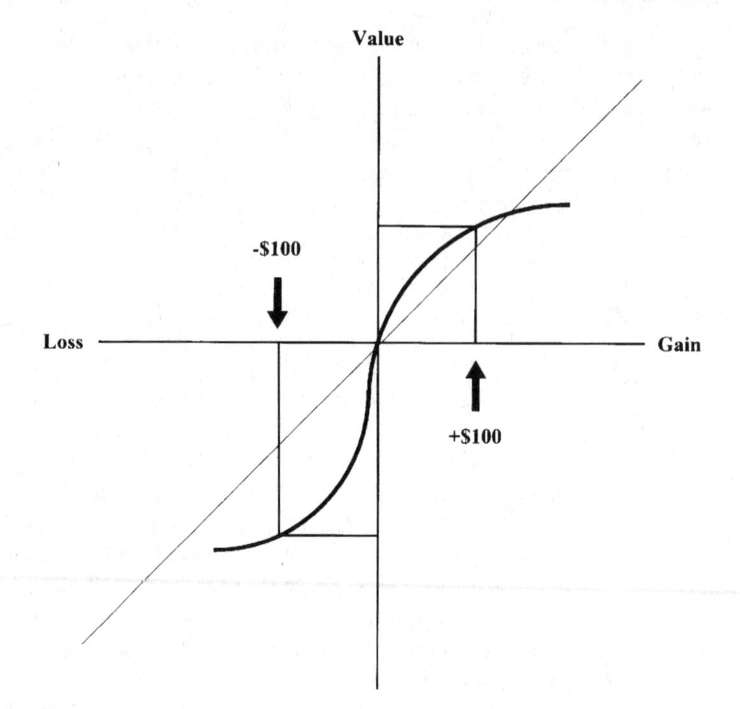

Fig. 6. A value function (Kahneman and Tversky, 1979)

may loom larger than the advantages that may accrue from the replacement.

Of importance to the manager is the insight that how a question is framed will affect the choice outcome. As an example we can take a problem quoted by Bazerman (1986, pp. 49–50)

> A large car manufacturer has recently been hit with a number of economic difficulties, and it appears that three plants need to be closed and 6000 employees laid off. The vice-president of production has been exploring alternative ways to avoid the crisis. She has developed two plans

Plan A: This plan will save one of the three plants and 2000 jobs.

Plan B: This plan has a one-third probability of saving all three plants and all 6000 jobs, but it has a two-thirds probability of saving no plants and no jobs.

When tested, most subjects (over 80%) chose Plan A. When, however, the choice was framed in a different way another result occurred.

Plan C: This plan will result in the loss of two of the three plants and 4000 jobs.

Plan D: This plan has a two-thirds probability of resulting in the loss of all three plants and all 6000 jobs but has a one-third probability of losing no plants and no jobs.

In this case over 80% chose Plan D.

This is a particularly good example of the effect of framing. Despite the fact that, in accordance with EU theory, all four plans are equally attractive the subjects were clearly biased towards either the sure thing or the gamble depending on whether the problem was framed in terms of gains or losses. This behaviour corresponds closely with the precepts of prospect theory and badly violates the principles of EU theory.

Managerial attitudes to risk

In expected utility theory both probability and outcome are important and risk is defined as a function of these two factors. In management circles, however, quite another view of risk is dominant. March and Shapira (1988) have reviewed the empirical evidence and come to the following conclusions.

- Outcomes that have very low probabilities associated with them tend to be ignored. Thus the catastrophic, high consequence, low probability events associated with natural disasters, major physical hazards or World Wars are not included in a manager's portfolio of concerns.

- Probability expressed in fuzzy verbal terms is preferred to numerical estimates which are thought to be gross oversimplifications of the complexity of reality.
- Risk is generally understood in terms of negative outcomes unassociated with probabilities. This is normally expressed as the amount of monetary loss possible if a certain decision is made.
- Taking risks is part of the managerial ideology. However, although risk taking is thought to be intrinsic to management, in real organizational life the manager does all in his or her powers to avoid or minimize risk. Most managers feel that risk can be controlled and reduced.
- Propensity to risk taking appears to depend more on circumstances than personality. For example, it is speculated that a decision maker achieving outcomes just above a target will be risk averse but those either well above or just below the target will give their attention more to opportunities than risks and as a consequence be inclined to risk taking.

The significance of these entrenched attitudes to the budding decision analyst are clear. Do not expect your boss to accept the results of the decision analysis exercise at its face value. Do not be surprised if your efforts are ignored.

Is decision analysis useful?

Despite the wide differences between the assumptions of EU theory and the reality of human judgement, decision analysis as a decision aid, has a useful role in managerial decision making. Decision analysis is a robust tool where circumstances allow the choice between discrete entities in situations where forecasting is important. Forecasting sales or yields are an important function of management and the ability to structure choice using decision analysis is often useful. The ability to update the forecast in an axiomatically rational manner is one of this method's strengths. Usually, however, only some of the decision variables can be modelled in the form of a decision tree. It is wise, therefore, to treat the results of decision analysis as one more piece of data to be pro-cessed judgementally alongside those other important factors that could not be suitably modelled.

Summary

- Decision analysis is the most common use of expected utility theory in management. It uses decision trees to display the interaction of the subjective probability estimates and monetary pay-offs of future actions. It is of particular use when forecasting is important in a decision.
- One of the drawbacks of decision analysis is that human beings handle probabilistic tasks poorly. It also has the disadvantage that it does not conform to the observed behaviour of real decision makers.

5

Multiattribute choice and other rules

This chapter concentrates on choice rules that do not use probabilities. As has been noted, probabilistic thinking is hard for most people and the difficulties are compounded by the volatility of modern life which makes forecasting so problematic. Luckily, we have a number of non-probabilistic techniques that may be of help to the decision maker.

Multiattribute choice

Out of the same body of research that nurtured decision analysis, other closely associated work in operations research, and research into the psychology of judgement, has come a number of techniques that do not use probabilities (von Winterfeldt and Edwards, 1986; Watson and Buede, 1987). These techniques model choice using the present state of nature rather than some forecast of its future state. This is the sort of decision that requires the choice between a number of existing objects such as sites for a dam or pieces of equipment, or perhaps the prioritization of a set of projects for funding.

In the three methods considered in this section the options to be compared are assumed to have a set of desirable attributes that can be scaled in some way. Thus, the decision maker can place each attribute of each alternative on the scale either directly or comparatively. But each option will have a number of attributes, some of which will score higher than those of other alternatives and some of which will score lower. Dominance created by any option producing a high score on all attributes is unlikely and, of course, would make the choice obvious. So how do we calculate a universal number (utility) that reflects our overall preference of one object over all others? In all three methodologies considered here the key is the elicitation, from the decision maker, of his or her numerically expressed preferences for each of the attributes. In other words, given the goal of the exercise, how much more (or less) important is one attribute than the others. Deriving these weights is the essential distinguishing characteristic between the methods. One method simply asks the decision maker to estimate relative importance, one gives the decision maker a set of paired choices, and the other asks the decision maker to make holistic choices to enable the weights to be back-calculated. In all cases an overall utility for each option can be calculated by multiplying the importance weights by the attribute score on the scale, and summing. Clearly, such methods are of most use when the problem attributes can be defined and scaled, and the decision maker's preferences measured, but the exact nature and number of the options for choice are unknown. Thus a model is produced that reflects the decision maker's preferences and can be applied to both present and future choices. The three methods reviewed draw on a report to the USA Air Force Human Resource Laboratory by Fast and Looper (1988). In each case we will consider the same case study of an air force promotion system. The aim is to design a model to aid promotion decisions based on six key attributes. The attributes for choice are

1. scores on a job knowledge test (JKT)
2. scores on a general organizational knowledge test (GOK)
3. time in service (TIS)
4. time in grade (TIG)
5. awards and decorations (AD)
6. an individual performance rating (IPR).

SMART (simple multiattribute rating technique)

Developed by Ward Edwards, one of the foundation researchers in expected utility theory, this is the simplest of the three methods (von Winterfeldt and Edwards, 1986).

The SMART procedure in the air force case is as follows.

Elicit single-attribute value functions

These are the scales for each attribute. All six scales go from 0 to 100 and may not be linear. For example, in the case of TIG, 0 = 0 months and 100 = 120 months with an S curve as the function shape. For attribute AD, 0 = no decoration, 100 = combination of Medal of Honour, Purple Heart and Silver Star. Other combinations of decorations would receive levels between these on the scale.

Elicit weights

The decision maker is asked to rank each of the attributes in relation to the goal. The top rank is given a score of 100. Each other attribute is then tested against this standard by asking the question 'how much more value would this attribute contribute if its score was increased to from zero to 100?'. This may be (for example) 80% or 20% of that contributed by the most important attribute. These weights are then normalized to sum to 1. Weights are often obtained in stages using a *value tree*. In our case JKT and GOK may be assumed to be attributes of Knowledge (KNOW) and TIS and TIG as attributes of time (TIME). The value tree would look like Fig. 7.

Thus, IPR, AD, TIME and KNOW are compared followed by comparison between TIG and TIS, and GOK and JKT. The final weights are obtained by multiplying through the tree.

Aggregate

The model can now be used to give a promotability value to each candidate. Each candidate's qualification is compared to the attribute scales and he or she is given a score. These scores are then multiplied by the appropriate weights and summed. One candidate is shown in Table 5.

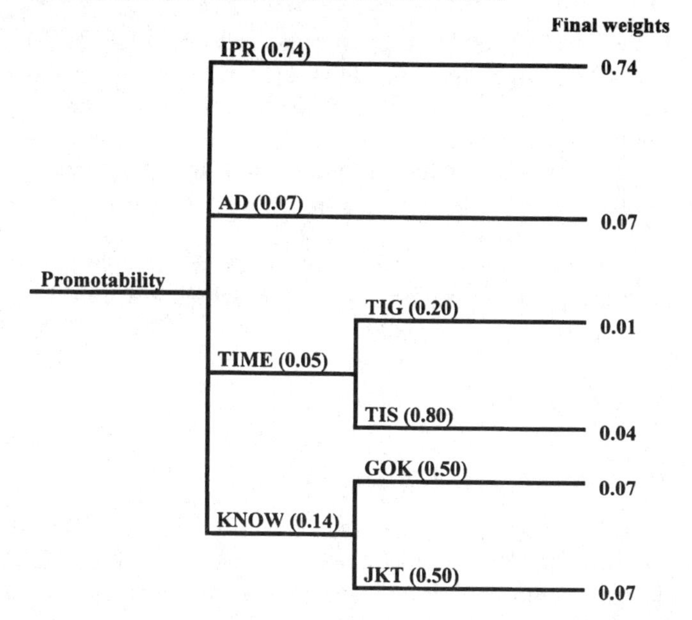

Fig. 7. Value Tree for promotability (Fast and Looper, 1988)

Table 5. One candidate's score (Fast and Looper, 1988)

Attribute	Candidate's profile	Score on the scales	Weights	Promotion values
JKT	50 pts	50	0·07	3·50
GOK	75 pts	75	0·07	5·25
TIS	60 months	50	0·04	2·00
TIG	12 months	25	0·01	0·25
AD	AF Commend	25	0·07	1·75
IPR	100 pts	80	0·74	59·20
				71·95

It is important to note that the attributes must be independent of each other. Movement on one attribute scale must not cause movement on any other. This is not easy to achieve in practice as most real attributes have some degree of interdependence. Check, therefore, for both comprehensiveness and independence and redraw the tree to maximize these virtues. Always use tests sensitively in complex cases.

The next two methods are not quite so easy to explain because computers are used to derive weights. Nevertheless I will do my best.

AHP (analytic hierarchy process)

The version of Saaty's (1980) analytic hierarchy process considered by Fast and Looper (1988) was that described by Hwang and Yoon (1981). The modelling process begins with the construction of a hierarchy structure rather like SMART's value tree but with the candidates (O_J) at the bottom (Fig. 8).

Elicit weights

Starting from the upper layer of attributes, the decision maker is asked to compare pairs of attributes. 'Which of the pair KNOW and TIME is more important? On a scale of 1 (equal) to 9 (very much more) is one more important than the other?' All combinations are compared and a matrix prepared. AHP software (using eigen values), then solves for the best fitting normalized weights given some degree of inconsistency in the ratios. The matrix for our example is shown in Table 6.

At the next level, as not more than two attributes exist at each node, the matrix is a simple 2 by 2 with no possible inconsistency. In this case the weight ratio of JKT to GOK is 1, giving relative weights of 0·50 for JKT and 0·50 for GOK. The same is done for TIS and TIG resulting in relative weights of 0·80 and 0·20.

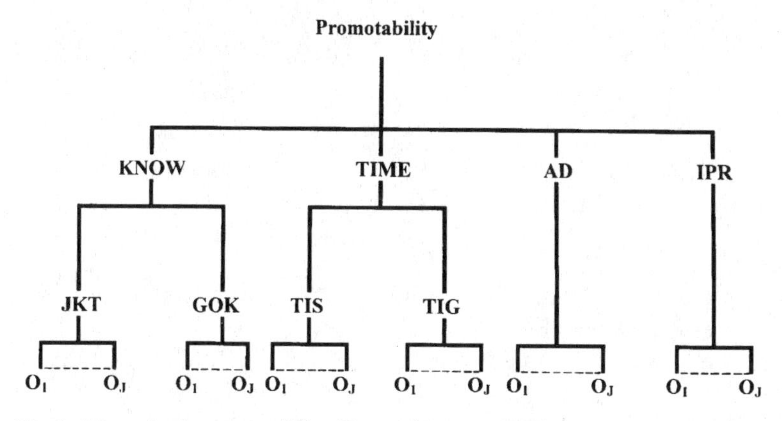

Fig. 8. Hierarchy for promotability (Fast and Looper, 1988)

Elicit preference scores

At this point we must have a list of candidates. In our example we will assume five. The decision maker will be asked, taking one pair of candidates at a time, 'which do you prefer in respect to the attribute under consideration?' (say JKT). We will go on to ask, 'on a scale of 1 to 9 (1 = indifference) how much more do you prefer one to the other?'. Again a matrix, like the one in Table 7, is formed and analysed using eigen values.

Table 6. A typical pairwise comparison matrix (Fast and Looper, 1988)

	KNOW	TIME	AD	IPR	Normalized WTs.
KNOW	1	3	2	1/9	0·13
TIME	1/3	1	1/2	1/9	0·06
AD	1/2	2	1	1/9	0·09
IPR	9	9	9	1	0·72
			Inconsistency score		0·054

Aggregate
In a similar way to SMART, the overall value of each candidate is arrived at by multiplying through the tree. Thus the overall value of candidate O_1 is

$$V(O_1) = (0{\cdot}13 \times 0{\cdot}50 \times 0{\cdot}33) + (0{\cdot}13 \times 0{\cdot}50 \times a) + (0{\cdot}06 \times 0{\cdot}8 \times b) + \cdots (0{\cdot}72 \times 1{\cdot}0 \times e)$$

where a, b, c, d, e are the candidate O_1's relative scores for GOK, TIS, TIG, AD and IPR.

It is clear that the major drawback of this method is that it compares batches of candidates against each other rather than against a scale. Thus, as each new candidate comes forward the process has to be repeated. It is possible, of course, when repeating the process that rank orders will change and a candidate's score will fluctuate. It follows that AHP is most useful when the number and nature of the alternatives are known.

Policy capturing

This method is derived from social judgement theory (Hammond *et al.*, 1975; Cooksey, 1995) originally as a research tool, but with some later applications in policy consulting (Parkin, 1993).

Table 7. Preference matrix for five candidates for JKT (Fast and Looper, 1988)

	O_1	O_2	O_3	O_4	O_5	Relative score for JKT
O_1	1	6	3	2	1	0·33
O_2	1/6	1	1/2	1/3	1/6	0·05
O_3	1/3	2	1	2	1/3	0·14
O_4	1/2	3	1/2	1	1/2	0·14
O_5	1	6	3	2	1	0·33
					Inconsistency score	0·034

Table 8. Typical candidate profile (Fast and Looper, 1988)

Applicant	
JKT	Score of 86
GOK	Score of 65
TIS	50 months
TIG	13 months
AD	Purple heart/ Airman's medal
IPR	135

Elicit weights and function forms

The decision maker is presented with a fictitious set of candidate profiles expressed in terms of the attributes (Table 8).

A statistically appropriate number of imaginary candidate profiles are presented and the decision maker is asked to score each candidate out of (for example) 100. Thirty or so candidates are often enough in non-research settings. These candidate profiles should be as variable as possible but, nevertheless, should not contain combinations of attributes that would not normally be found in the real world.

The scores for all of the profiles are then analysed using multiple regression software, with the attributes as independent variables and the decision maker's judgement as the dependent variable. Relative weights for each attribute are calculated by dividing each beta weight in the regression equation by the sum of the beta weights. The function form of an attribute (holding all other attributes constant) can also be calculated, together with a multiple correlation coefficient as a consistency check. The weights, correlation matrix, function forms and coefficient represent the decision maker's policy.

Using the policy

The multiple regression equation, constructed using the decision maker's holistic judgements, can now be used to assign a score to all future candidates. The profile of each candidate is assembled using the attributes, 'bootstrapped' into the decision maker's regression equation, and that candidate's score out of 100 is automatically calculated.

The major disadvantage of policy capturing lies in its inability to handle more than about seven attributes at a time. This is because human beings have difficulty in integrating large numbers of pieces of information. Our natural response to a larger task would be to disregard some of the attributes altogether. In my own experience I have found policy capturing particularly useful when the problem is highly salient to the decision maker, the number of attributes is small but conflicting, and the bootstrapping of future cases is required. If the number of attributes is large, I have found that SMART is highly acceptable to decision makers because of its transparency and ease of calculation. Paradoxically, despite its simplicity, some clients want the final SMART value tree loaded on to a computer for future use.

Decisions without probabilities

Sometimes forecasting is very difficult but decisions have to be made. In the circumstances where discrete pay-offs can be estimated but their probability cannot, there are a number of decision rules that could be applied. These rules will be illustrated using the following example adapted from Moore and Thomas (1976).

Your consulting company has expanded to full capacity within the confines of your present accommodation. You do not want to move because the likelihood of sub-letting is small. You have consulting proposals submitted which will require extra resources if won. If they are not won your work load will drop off slightly.

Table 9. Pay-off matrix

	Do not get projects	Get projects
CAD	$100 000	$250 000
Overtime	$120 000	$220 000
Do nothing	$120 000	$150 000

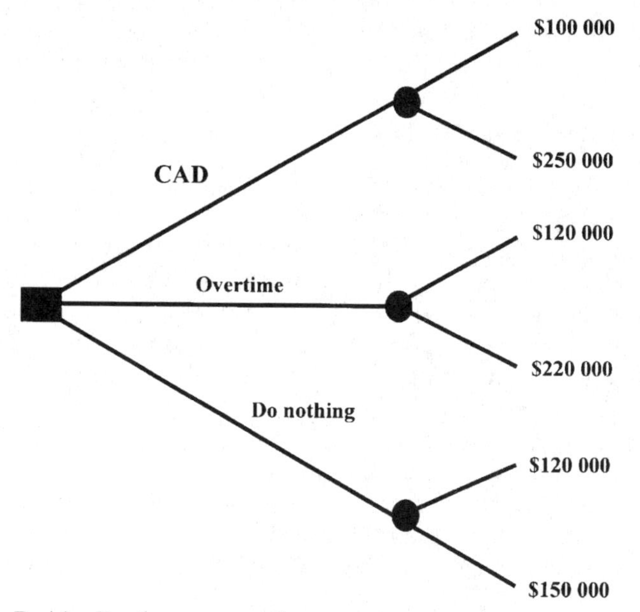

Fig. 9. Decision Tree for resource options

These resources can be provided by introducing overtime for engineers (and setting a bad precedent) or by investing in better computer aided design (CAD) equipment. The pay-off (net income) matrix and decision tree are shown in Table 9 and Fig. 9.

You admit that you do not know the chances of getting the projects. You try a number of rules (Mullen and Roth, 1991).

Maximin rule

This is for the pessimists. Using this rule you will assume no projects and pick the maximum of the low pay-offs—$120 000 for 'do nothing' or 'overtime'.

Maximax rule

For optimists. Choose the option that maximizes the maximum pay-off by buying the CAD system.

Greatest average pay-off rule
For the compromisers. The average pay-off for CAD is $175 000, for overtime it is $170 000, and for do nothing it is $135 000. Therefore, choose the CAD option.

Minimax regret rule
This rule is designed to minimize the regret that occurs when our judgement is not perfect. Thus, if we choose CAD and do not get the projects we feel a degree of regret expressed as the difference between the maximum pay-off and the CAD pay-off in the 'do not get projects' column. With CAD we would feel no regret if we did get the project. However, if we choose one of the other two options the regret will be associated with the 'get projects' column. We should choose the option that minimizes the maximum regret possible. A pay-off matrix is shown in Table 10. Under these circumstances because the minimin maximum regret is the $20 000 (used to purchase CAD) this seems to be the choice using this rule.

Approval voting rule
This involves finding the average pay-off under 'do not get projects' and under 'get projects' and mark each option as acceptable if its pay-off is greater than the average. Select the option with the greatest number of acceptables. Thus the 'do not get projects' average is $113 000 and so overtime and do nothing are acceptable. The average for 'get project' is $206 000, and overtime and CAD are satisfactory. Our choice should be to prepare for overtime, as this attracted two satisfactories.

Table 10. Regret matrix

	Do not get projects	Get projects	Max. regret
CAD	$100 000	$250 000	$120 000–$100 000
Overtime	$120 000	$220 000	$250 000–$220 000
Do nothing	$120 000	$150 000	$250 000–$150 000

Table 11. Ranking matrix

	Do not get projects	Get projects	Average rank
CAD	$100 000 (1)	$250 000 (5)	(3)
Overtime	$120 000 (2)	$220 000 (4)	(3)
Do nothing	$120 000 (2)	$150 000 (3)	(2·5)

Ranking rule
Rank each pay-off from the least to the greatest. Find the average of the ranks for each option and choose the maximum (Table 11). Using this rule we should choose either to install CAD or to introduce overtime.

How do we choose the rule?

Mullen and Roth (1991) have suggested some principles to guide the decision maker when faced with decision problems that would normally be approached using decision analysis but where the probabilities are difficult to estimate.

1. If one pay-off is significantly worse than the others and the others are reasonably grouped, use the *maximin rule.*
2. If one pay-off is significantly better than the others, and the others are reasonably grouped, use the *maximax rule.*
3. If you have reason to believe that the probabilities, if known, would not be significantly different from each other and the use of maximin or maximax are not indicated, use the *greatest average pay-off rule.*
4. In the same circumstances to number 3 above, and where the pay-offs can only be expressed as ranks, use the *ranking rule.*
5. In the same circumstances that would call for the use of the maximin rule (number 1 above) but where you feel that feelings of regret would result from an incorrect choice, use the *minimax regret rule.*

6. Where the maximin and maximax rules are inappropriate and where some common sense definition of acceptability can be stated (such as above average) use the *approval voting rule.*

The mention of common sense in No. 6 is appropriate to all use of decision rules. In the example of the CAD versus overtime choice, it is unlikely that any analysis would be the final word on the issue. CAD systems are a long term investment, and introducing overtime payments for engineers may be unsuitable in the long term. Usually, you will find that many factors will impinge on the final decision that cannot be modelled in your analysis. I suggest, therefore, that you always treat the results of analysis as just another source of data for a holistic judgement.

Summary

- Multiattribute methods examine the present state of nature rather than some forecast state. The three methods examined all numerically scale the attributes of the problem but vary in their methods of weighting the attributes and the preferred form of the combination of weights and scales.
- Where pay-offs can be estimated, but their probabilities cannot, various rational rules may be applied to produce a sensible choice.

6

Engineering and social risk analysis

This chapter uses physical risk as a means of demonstrating the way that engineering management decision making can be negatively influenced by our tendency to model problems in purely technical terms. Engineering risk analysis is an excellent example of the way that our engineering values and methods can blind us to other world views. It demonstrates that, when engineers are required to make apparently technical decisions, they may mistakenly ignore other important social constructions of events.

The technical construction of risk

We engineers are a conservative lot. We do not like risk, and as a consequence we have developed a culture that is guided by standards, codes of practice and specifications to ensure that what we create is safe as well as functional. We have a philosophy of safety, and our traditional language, using phrases like *safety factor*, reflects this philosophy. Only quite recently have design codes talked of *failure probabilities*—shamefully admitting our covert, unacknowledged lack of perfection. Despite our rules, things do break, collapse and blow up. Engineers have faced the truth that unanticipated events can spoil their plans, that we do not know the full nature of the applied loads, the characteristics of the materials, or the consequences of all the possible interactions—that, in reality, our world is probabilistic. The point of emphasis varies across engineering. In some cases it is the extremes of loading arising from natural forces that has attracted probabilistic attention. In others, it is the short and long term carrying capacity of new materials, and in the cases of the nuclear and petrochemical industries, it is the potential catastrophes which may result from the failure of complex, tightly coupled systems. In civil and mechanical engineering the emphasis has been on failure probabilities (reliability) with the consequences being only implicitly acknowledged. However, in those industries dependent on new untested technology, analysis has been forced to explicitly take account of the potential for death or injury. As a consequence, risk analysis in these industries has adopted a form of expected utility theory to link failure probabilities with their outcomes. One such industry, petrochemicals, illustrates the nature of our taken-for-granted technical risk rationality and its consequences for decision making.

Probabilistic risk analysis

Petrochemical plants and transportation vehicles often contain material that if released, would explode, burn or gas people. The danger may be confined to the immediate vicinity of the container or may involve the surrounding built-up area. Casualties can be high. In an effort to design petrochemical facilities in such a way that they can be considered safe, engineers have made liberal use of probabilistic risk analysis. At the core of this method is a requirement to evaluate the probability of a hazardous event and the likely consequences for the surrounding population. The units used are normally estimated frequencies (F) and number of deaths (N), and risk is defined as a function of these factors (Philipson and Napadensky, 1982). Thus, as in expected utility theory, a small probability of a major consequence is deemed equivalent to a high probability of a small consequence—although the relationship is not necessarily linear. The risk analysis typically contains the following steps (European Federation of Chemical Engineers, 1985; NATO, 1984).

Quantify hazardous event probabilities

The estimated frequency of a hazardous event in the petrochemical industry cannot often be quantified directly from records of previous catastrophes because of their rarity and circumstantial dependence. Hence attempts are made to predict rare large events from a knowledge of the frequency of more common small failures. The *fault tree* is used in large complex plants and links the uncommon top event to the failure of basic causal events (like a sticking valve), using Boolean logic. Thus, a knowledge of the past history of failure of the subunits (such as valves), enables the calculation of the probability of failure of a critical part of the plant likely to precipitate a hazardous event. Human error is often included in the statistics of equipment failure, but nevertheless, remains a major source of estimation error (Cox, 1982). *Event trees* are designed to predict the probability of certain types of outcomes if the nature of the hazardous event is known. Thus, if the probability of a dangerous fault or accident can be estimated from the fault tree or historical records, then an event tree can be used to calculate the probability that this will result in various types of consequences.

As an example, we can consider the problem of transporting LPG through city streets (Dryden and Gawecki, 1987). Firstly, the tanker route is divided up into segments and the probability of a tanker crash calculated using transportation accident statistics. This is the estimated frequency of the top event. Given the occurrence of this hazardous event, the probability of a range of consequences can be calculated using data from a fault tree analysis or historical records. A typical example of an event tree for an LPG tanker accident is shown in Fig. 10. It can be seen that the probabilities of really dangerous events such as a large torch fire, a fireball, or an unconfined vapour cloud explosion (UVCE) is small, but the consequences in a built-up area could be grave.

Quantify the consequences

A knowledge of the physics of the hazardous event enables an estimate to be made of the likely casualties. For example, the crash of an 8 tonne LPG truck, which resulted in an UVCE, would almost

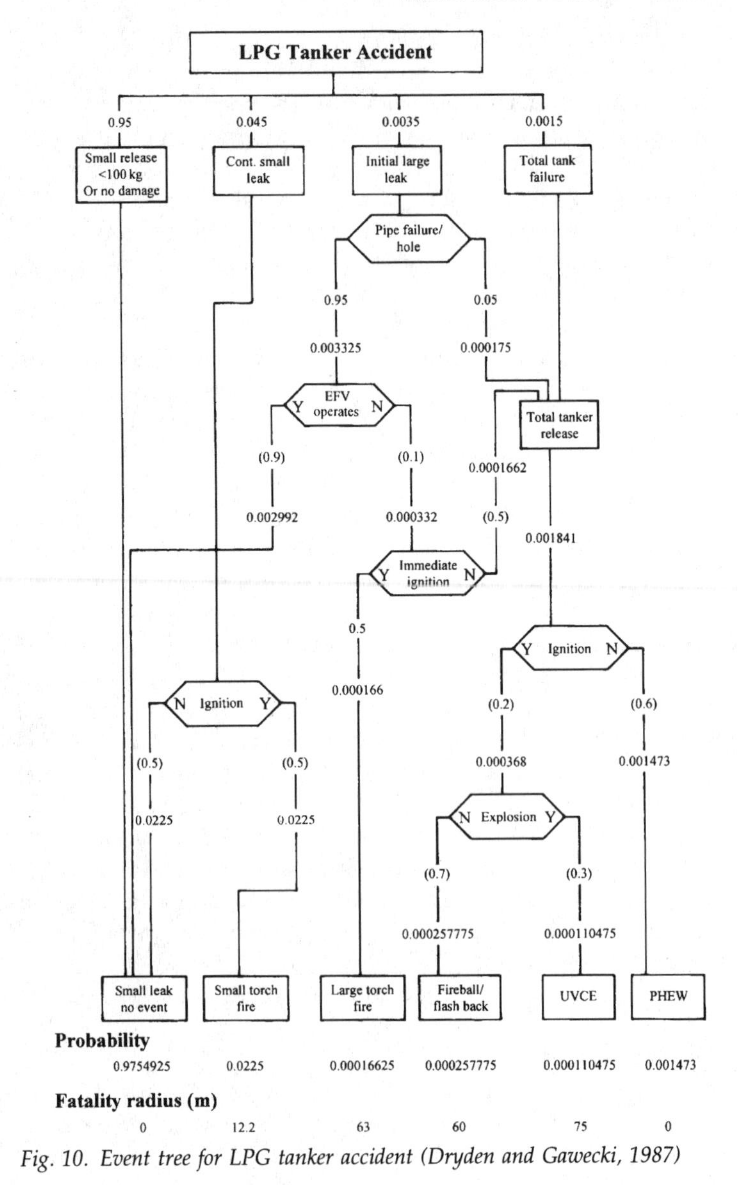

	Small leak no event	Small torch fire	Large torch fire	Fireball/ flash back	UVCE	PHEW
Probability	0.9754925	0.0225	0.00016625	0.000257775	0.000110475	0.001473
Fatality radius (m)	0	12.2	63	60	75	0

Fig. 10. Event tree for LPG tanker accident (Dryden and Gawecki, 1987)

certainly kill exposed people standing within 85 m of the explosion and cause injury to those within 170 m (Pitblado, 1984). Thus, depending on the time of day and the use of the street, the consequences, expressed in terms of death can be calculated for each route segment.

Quantify the risk

- *Individual risk.* This is often expressed as the probability that an individual within an exposed group will be killed in the next year because of a hazardous event. This is then compared to some acceptable risk criteria such as an increased probability of death of 1×10^{-6} per year.
- *Societal risk.* This is often expressed as the estimated frequency of hazardous events with consequences above N, where N is the number of casualties. The acceptability of these frequencies is checked against some standard such as the one shown in Fig. 11. This was adopted for the analysis of risks associated with the transport of dangerous goods in Sydney (Dryden and Gawecki, 1987).

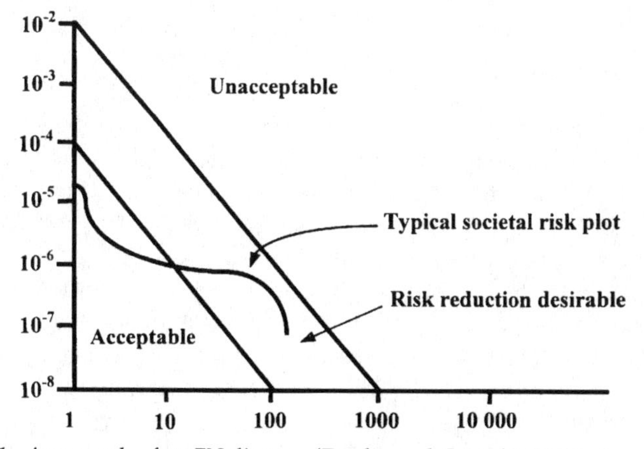

Fig. 11. An example of an FN diagram (Dryden and Gawecki, 1987)

Often these standards and those for individual risk were originally deduced from the statistics of natural hazards or common life hazards that appear to be acceptable to the general public.

Risk analysis as an engineering decision technique appears rational but has nevertheless attracted much technical criticism (Whittaker, 1991). Firstly, the role of human error is largely ignored. Although average, small mistakes are accounted for in the statistics of equipment failure, the incidence of gross failure is ignored. Yet tragic failures such as Bhopal, Challenger, Chernobyl and even the King's Cross Underground fire can be sheeted home to human errors that could not have been anticipated by risk analysis (Reason, 1990). Secondly, the accumulating historical evidence of failures in the nuclear, aerospace and petrochemical industries indicate that fault tree predictions grossly underestimate the probability of hazardous events. This is partly due to the dominance of unanticipated failure interactions in complex facilities (Whittaker, 1991). Yet these technical doubts pale before the torrent of criticism that has been voiced by social scientists. The essence of this critique is that the technical rationality of risk analysis and, indeed, applied expected utility theory in general, does not match the reality of humanity's real sense of risk.

Cultures of risk

This chapter is about how we make decisions concerning engineering hazards. This would appear to be a universal theme independent of cultural factors. Such a view is unsustainable if we imagine ourselves in the position of (for example) a peasant in Bangladesh. As a poor farmer we expose ourselves to hazards that would be unthinkable in the rich northern countries. We may face our dangers and hardships with fatalism or bitterness but face them we must—or die. The fear of starvation eclipses all other fears. Other fears may also dominate the fear of technological risk, such as those generated by deeply held religious beliefs. Indeed, from the point of view of much of the world's population, always living close to poverty, the discussion of technological risk must seem bizarre or frivolous. Yet in our industrialized cultures, an obsession with technological risk is pervasive and growing. We live in what Beck (1992) calls a 'risk society' where, to some perceptible degree, the culture of techno–economic development is losing legitimacy to be replaced by a risk culture. As commentators such as Wartofsky (1986) have pointed out, this is associated with an increasing tension between our sense that we should control our own personal destinies and the apparently increasing imposition on us of technical hazards. Risk is salient to us because we see that such hazards can be avoided if we take collective action. Unlike the helpless poor of the world we feel empowered to demand the mitigation of imposed hazards. Whatever risks we are willing to take in our voluntary leisure time, we feel indignation when others impose risks upon us or our families. And we are self confident enough to take political action in an attempt to frustrate such impositions.

Although industrial cultures are often more sensitive to technological risks than other cultures, there are groups within such societies that take a more robust view. Generally speaking, those involved in or committed to technology or business have less of a sense of risk than those who do not (Buss *et al.*, 1986; Buss and Craik, 1983; Deluca *et al.*, 1986). This is understandable, as making a living is an important motivator for an engineer or factory worker. It is so much easier for a teacher to worry about the pollution from a power station than the shift engineer with three children to bring up. Indeed, a group's attitude to risk can often be predicted from a knowledge of their attitude towards, and trust in, industry, government agencies and science—the key institutions of modernity (Bastide *et al.*, 1989; Otway and von Winterfeldt, 1982; Bord and O'Conner, 1990).

The psychology of risk

The application of expected utility theory of risk does not correspond to most peoples' judgement of riskiness. Some low probability, high consequence risks are feared much more than more common risks with potentially high impacts (von Winterfeldt *et al.*, 1981; Kunreuther, 1992). Probabilities are irrelevant because people either discount them intuitively—'it can't happen to me' or they focus on the consequences. Why is this?

The work of Slovic *et al.* (1986) and Slovic (1992) is central to this question. Using elaborate questionnaires and complex statistical analysis techniques, these researchers have mapped the attitudes of diverse groups in the USA to risk, which other researchers have replicated in other countries.

The key finding is that experts and non-experts think about risk in very different ways. Risk experts rate the degree of risk associated with a technology with the likely casualties. Non-experts could make casualty estimates similar to those of the experts, and did rate this factor as important. Nevertheless, non-experts were distinguished by a broader concept of risk. Factors such as the catastrophe potential, controllability, and threat to future generations were important. Analysis found that two clusters of factors were related and important in any assessment as to whether a technology was risky. The first cluster indicated how unfamiliar, unknown or delayed were the effects and the second cluster indicated how uncontrollable, dreaded, catastrophic, fatal, unfair, dangerous to future generations, increasing, and involuntary were the effects. An increase on one or both of these scales indicated increased fear of technology. Technologies associated with DNA, radioactivity, toxic chemicals were high on the dread scale. Many technologies associated with everyday life and sport that produced significant numbers of deaths were low on these risk scales. Most people felt that the levels of technological risk were too high and should be subject to more regulatory control—particularly those high on the dread scale. Other work by Stallen and Thomas (1984), in Holland, indicated that a feeling of personal control was important. In line with related stress research, they concluded that people cope with threats much better if they feel that their destiny is in their own hands.

Amplification

Early work on the psychology of risk indicated that small incidents associated with technologies high on the dread and unknown scales, produced strong negative reactions (Slovic, 1992). The nuclear accident at Three Mile Island demonstrated this principle. Although no fatalities resulted from the leakage of radiation, the social effects were very great. People saw the accident as a signal or warning that much more catastrophic accidents could just as easily occur in that industry. The resulting political fallout was fatal for US nuclear technology.

Kasperson *et al.* (1988) have put together a model to explain why technically minor incidents often cause strong public reactions. This model uses the analogy of a radio signal to account for the way the nature of risk changes as it moves from the domain of experts into the public sphere. The risk event gives out certain signals which are interpreted in different ways by different groups and translated into role-dependent languages. Thus, the original signal is decoded, filtered, subject to cognitive manipulation and value loaded in a variety of ways. These processes may serve to radically amplify the signal as the receptor groups pass the message on. Amplification may result from the volume of news surrounding the event, the conflict generated between experts and its real or potential, symbolic and dramatic nature. The social consequences such as litigation, investor flight and community mobilization may be seen to be a function of this social amplification rather than the risk event itself.

The reader may note the similarity between this model and the Brunswik lens model discussed in the next chapter. In a Brunswikean sense the perceptual cues associated with the event are selected and integrated in different ways by different people. An engineer will tend to look at the physical damage, its causes and casualties. The surrounding population will notice those cues that indicate the long term safety of the facility and the worth of the safety features. Officials will be concerned with the aspects that contravene regulations or indicate that regulations may have to be modified. The press, of course, are looking for sensational cues that will sell newspapers. If they cannot find them at the site of the incident they can always wheel out a hired expert to point the finger of blame and warn of the potential for catastrophe. The owner will tend to move from a pragmatic mood confined to the physical and production consequences, through an increasingly alarmed period where the publicity and conflict surrounding the plant seems to be getting out of hand, to the astonished realization that resulting commercial damage was unrelated to the engineering damage.

Stigma

An important social factor in the way non-specialists react to risky technology is the phenomenon of stigmatization. For example, waste disposal sites may be judged as repellent, ugly, upsetting and disruptive as well as dangerous. These are all negative stigmata associated with such facilities which are easily passed on to associated entities. For example, Slovic (1992) cites the example of dairy cows that became slightly contaminated by polychlorinated biphenyl (PCB). Even though the dairy was within health limits the stigma associated with PCB was sufficient to cause consumer flight. Whole cities or states can become stigmatized by an association with a source of fear. Slovic describes a study to look at the impact of locating a nuclear waste storage facility in the Yucca Mountains in Nevada. Using out-of-state subjects, the team was able to demonstrate that the stigma of such a site was sufficiently strong to have a potential effect on the attractiveness of Nevada, and Las Vegas in particular, as a tourist destination. The potential for the stigmatization of the whole state was such that any further investigatory work in the Yucca Mountains was actively resisted by the State Government (Slovic, 1992).

From an engineer's point of view, we must face the possibility that the phenomenon of stigmatization has permanently affected the public's view of certain industrial sectors. For example, in the word association tests used by Slovic and the consulting team, the most frequent word associated with chemicals was 'dangerous' or closely related terms such as toxic, hazardous, poison, or deadly. Beneficial uses of chemical were rarely mentioned.

Managing hazardous decisions

> It is increasingly apparent that the engineering sciences face a *historic turning point*: they can continue to think and work in the worn-out ways of the nineteenth century. Then they will confuse the problems of the risk society with those of early industrial society. Or they can face the challenges of a genuine, preventive management of risks. Then they must rethink and change their own conceptions of rationality, knowledge and practice, as well as the institutional structures in which these are put to work (Beck, 1992; p. 71).

The ability of an engineering manager to successfully manage decisions involving risk is limited by the magnitude of the social forces associated with the risk society. The sensitivity of the public to risk will increase despite the efforts of engineers to mitigate the negative effects of technology. Moreover, there is a danger that the whole of the engineering profession will be stigmatized in the process, rendering engineers a liability in decision making. Nevertheless, certain fundamentals are important when making decisions that involve a risk to life.

1. Do your risk analysis carefully. Your professional imperatives demand that you should be sure that all practical precautions have been taken, and that the analysis is robust enough to withstand peer group scrutiny.
2. Analyse the potential public reaction to your decision using a knowledge of the psychology of risk and the effects of amplification and stigmatization.
3. Give these social factors equal weight to the risk analysis and make your judgement holistically.

4. Use group or strategic decision management techniques to coordinate your efforts.
5. Communicate with all parties in an open and honest fashion, remembering that trust is the key to success.
6. Communicate in terms of consequences not probabilities.
7. Emphasize safety not risk; in particular, the safety record of your profession, its ethic of public safety, and the trust, built up over hundreds of years, that engineers are the protectors of the community.

Summary

- Engineers often structure physical risk as a function of the probability of a hazard and its negative consequences.
- The veracity of the technical analysis of physical risk is often degraded because of the difficulties of estimating event probabilities and an ignorance of human error.
- The general public do not use the same rationality as engineers. Hazardous technology is often feared for cultural and psychological reasons unrelated to casualties or probabilities. For this reason, decision making in this field must take account of both the technical and social construction of risk.

Part III

Holistic decision making

Three core chapters are included in this part of the book that describe what is known about the psychology and behaviour of decision makers. Chapter 7 examines what is known about the psychology of the decision process, with a particular emphasis on the act of judgement. Guided by organizational goals, the decision maker first responds to a problem or opportunity by thinking about the issues and potential solutions using available data. The data selected for use at this stage depends very much on the personal and role characteristics of the person making the judgement. This data is integrated using a combination of analytical and intuitive strategies and a judgement is made. The judgement may require a choice between alternative solutions or merely a judgement concerning the suitability of one solution. In an organizational setting, this judgement usually results in a decision, which in turn, generates action. Both the selection of the data (called cues in psychology) and their integration in the human brain are heavily influenced by a range of psychological biases, personality variables, social values and environmental influences.

Chapter 8 is more behavioural in emphasis, relying more on research into the decision behaviour of individuals working in organizations. The popular fully rational view of decision making assumes that we have an agreed set of goals, comprehensive information and the ability to optimize the solution. In practice, however, managers are subject to the normal cognitive limitations of human beings, are short of time for data search and analysis, and subject to organizational rules and norms of behaviour. As a consequence, real management decisions often fall short of the reasoned choice ideal. Research has indicated that models of organizational decision making can be devised that more faithfully reflect the true behaviour of effective managers. This research enables us to make some recommendations concerning the management of individual decisions which are not dependent on analytical aids, such as those discussed in Part II, but are based on an approximation to the holistic judgemental qualities of real managers.

Chapter 9 reviews what is known about group and team psychology and behaviour. This is an important matter because of the prevalence of the view among managers that teams are the ideal forum for decision making. It is certainly true that group decision making has advantages—particularly if the matter is complex and requires a range of expertise, or the democratic nature of the procedure enhances commitment. However, the disadvantages of team decision making, which seem to resolve around our human propensity to conform are also discussed. Nevertheless, despite these doubts, it is possible to indicate management strategies that exploit the advantages while avoiding the psychological traps.

7

Cognitive processes

This chapter is about the psychology of judgement and decision making. It concentrates on what we know about how our minds (our cognitions) react to problematic situations that require a considered and active response. In other words, what are the cognitive processes that guide our responses? Of course, the answer to this question must be constrained by our ignorance of brain functions and their interaction with the world. As this knowledge is, as yet, unavailable to humankind we must make do with what can be deduced from observation and testing, and the theories that model those observations. However, before embarking on our survey, it is necessary to frame the terms of enquiry using the simple model of the decision process shown in Fig. 12.

The holistic decision process

Figure 12 indicates that the process progresses from the recognition of a problem to some form of remedial action using cognitive steps we have called thought, judgement and decision. This is, of course, an ideal model designed to explain situations where the higher functions of the brain are involved rather than the motor reactions that seem to instantly jump from problem recognition to action. Figure 12 also indicates some of the influences that ensure that at all stages each human being will react in a different way to all others. Without an acknowledgement of this inconvenient fact, it

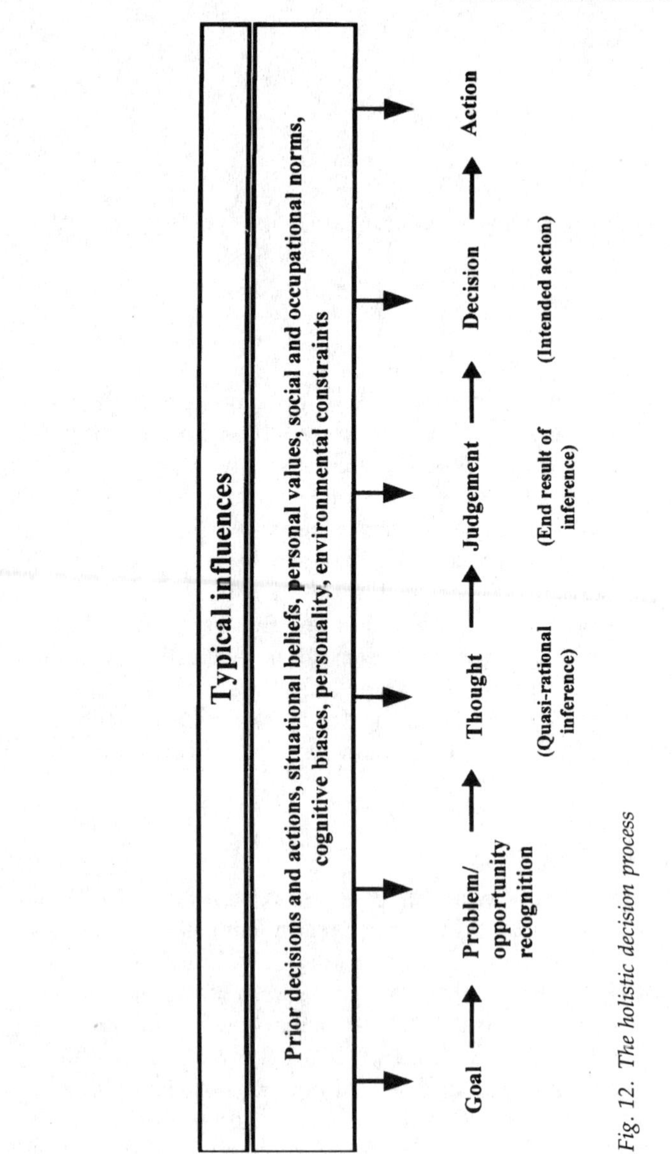

Fig. 12. The holistic decision process

would be pointless bothering to write about the decision process at all. All we would need to know is the nature of the problem and the goals of action, and the decision process would be somewhat similar for everyone. Almost as predictable as a machine. But, regrettably for us engineers, human beings are not only goal driven (and, therefore, potentially predictable) but are shaped by prior decisions, beliefs, personalities, values and social norms in a complex and dynamic flux of cognitive and physical activity. Predictability is therefore heavily constrained. However, for the student of decision making all is not lost in confusion. Some combinations of these influences and the cognitive processes depicted in Fig. 12 are fairly common and will be described in this chapter. We will begin with an explanation of the terms used in the central axis of Fig. 12.

Goals

Goals are the motivator in the decision process. They express human intentionality in concrete form. We have personal goals, social goals and, of most importance to our project, we have corporate goals. A large part of being a manager is the attempt, through legitimate means, to achieve previously agreed organizational goals. Often the achievement of organizational goals requires managers to make decisions, and all decisions and the resulting action should be directed towards those organizationally sanctioned goals. Moreover, decision making can be triggered by the inability to achieve goals. We can see then, that goals represent important constructs that shape the decision process. However, as has been noted in chapter 2, an organization may be the host to a portfolio of contradictory goals. One division's goals may conflict with another, and goal achievement may frustrate the achievement of higher goals. This implies that decisions in an organization may not fit together in a fashion agreeable to our engineering sense of order. So be it.

Problem or opportunity recognition

Situations do not pop up nicely labelled 'next problem'. Problematic situations, often created by others or resulting from our previous actions, are recognized as such when it is clear that our intentionality is about to be frustrated. Some problem is standing in the way of the achievement of a goal. It is the ambition to achieve a goal that creates the problem. But you may say, what about a house fire?—that is a problem to be solved that has not been created by the inability to achieve a goal. Well, a house fire is not a problem in itself. Spectators do not feel obliged to take action. Only the firemen are required to model the situation as a problem because they have a professional goal of a fireless neighbourhood, and this fire is standing in the way of its achievement. Other professional goals associated with the protection of life and property will mould the particular decisions to be made and the actions to be taken.

We do not, of course, require a problem for a decision. Often decisions are triggered by an opportunity to achieve a goal. To take a trivial example, the opportunity of a break in a meeting may cause you to decide to go and get a cup of coffee. At the strategic level, the opportunity to buy another company may trigger a whole series of decisions motivated by the corporate goal to penetrate a particular market.

Thought and judgement

Although these functions are shown separately in Fig. 12 they are only arbitrary, but useful, labels associated with one act of cognition. Judgement is defined as the end result of the quasi-rational search/interference process we have called thought. We are now talking about the modelling abilities of the brain and, as a consequence, we will have to concentrate on what is done rather than how it is achieved. The secrets of brain function are not at present open to review.

According to Egon Brunswik's (1952) influential theory of perception, the process may be modelled using the analogy of a lens. Fig. 13 is a diagram that shows, on the left, the problem, opportunity or solution we are thinking about. Part of this problematic situation has characteristics which may be reliably measured and agreed, together with other less physical characteristics that are less reliably reproducible across observers. For example, a problem or opportunity may be described on paper and have a physical presence so that its boundaries and content can be agreed. It will, however, have other less tangible characteristics to do with its place within a situation or its relationships to other, perhaps social, factors. We do not have to be philosophers to realise that it is difficult, perhaps impossible, to comprehensively define the essence of an object. It is even more difficult to comprehensively see a problematic situation—the boundaries and contents are fuzzy. Indeed, what may be a problem to one may not seem so to another—the situation is to a degree socially constructed. But even if we have consensus about the general characteristics of the situation, we cannot be sure that we will all agree about the details.

Brunswik (1952) contends that what we can perceive is a lens of cues derived from the object of contemplation. Thus, what we perceive is a model of the situation made up of these cues. Naturally, this model must be assumed to be an imperfect representation of the 'real' situation. Certainly, different observers have demonstrated that we all tend to perceive different cues given apparently the same stimulus (Hammond *et al.*, 1975; Cooksey, 1995). Cues that may be measurable on an artefact may be deemed 'objective' (Brunswik, 1952) and are often seen by most observers. Others may be situational or cultural and exclusive to only some observers. For example, a hill may manifest itself to a geologist as a remnant of a dyke, to an engineer as an obstacle, and to an aboriginal as a sacred site. Most organizational situations do not share this potential diversity of interpretations but, nevertheless, an engineer and an accountant may see quite different problems in the same situation. In terms of the lens model, each observer is perceiving a different set of cues. Some may be common and some exclusive.

The body of empirical work derived from Brunswik's psychol-

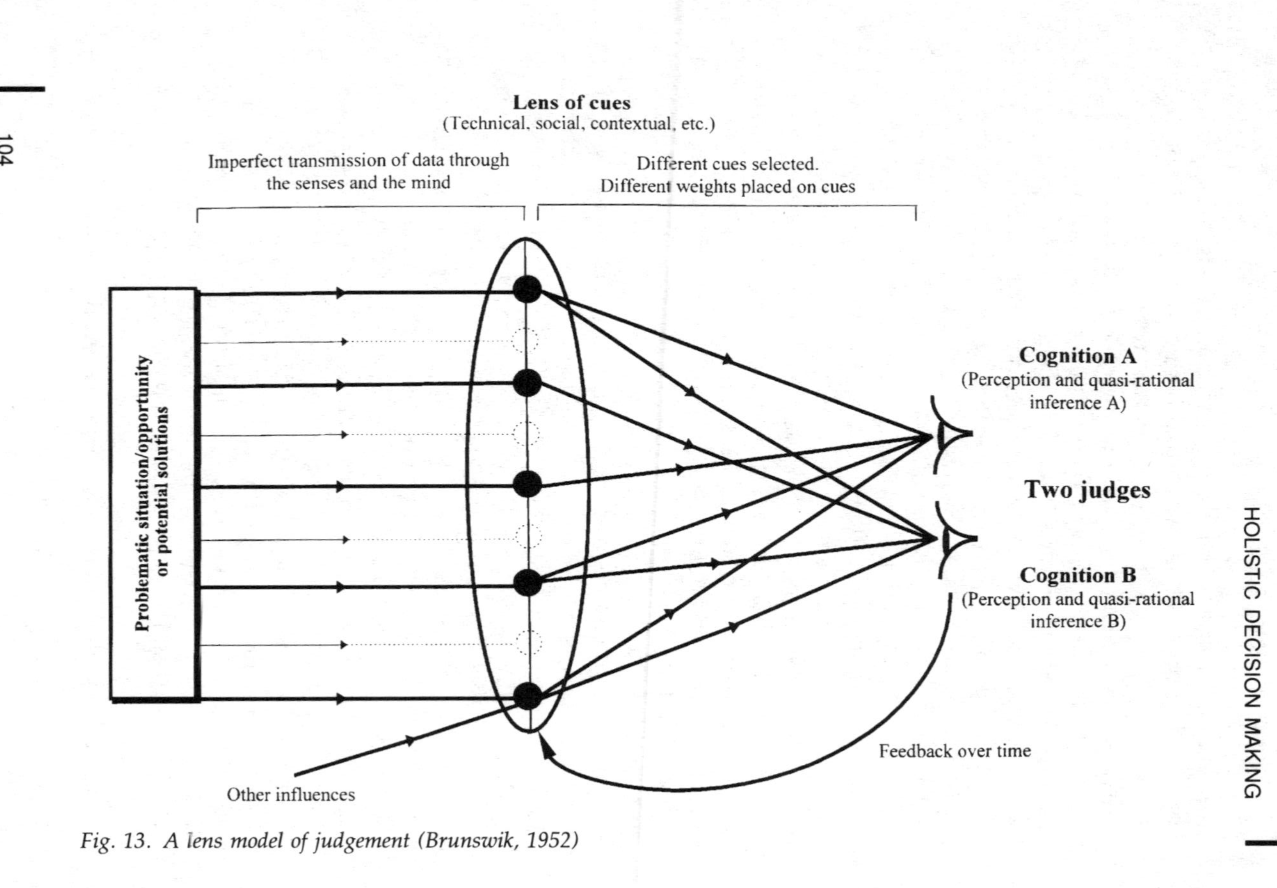

Fig. 13. A lens model of judgement (Brunswik, 1952)

ogy, known as social judgement theory (Hammond *et al.*, 1975; Cooksey, 1995) has measured other differences in the way we each use these cues. The cues must be subject to inference to enable us to come to a judgement about their meaning. This process is not necessarily the embodiment of reason (Hammond, 1993). It is in fact, probably only ever quasi-rational in its nature and the use of the word 'thought' may be misleading because of its association with analytical reasoning. Brunswik (1952) considers the perception of these cues to be quasi-rational because perception is an analogous process to reasoning but 'more primitive in its organization but vested with the same purpose' (p. 682). Hammond *et al.* (1983) have interpreted quasi-rationality as a compromise between intuitive and analytical cognition. Fig. 14 shows the nature of intuition and analytical thinking in their pure form (Cooksey, 1995). Managers operate in the quasi-rational mode between these extremes. Thus the perception of the cues may be more or less analytical or intuitive depending on the nature of the task for judgement and its complexity. Simple, especially numerical, cue sets are more likely to stimulate analysis, and the complexity of ambiguous cues are more likely to require a more intuitive approach. In real situations, as cues are presented in different forms, an individual may slide back and forth between intuitive and analytical modes (Hammond, 1993).

The quasi-rationality of perception means that the same judge often finds it difficult to explain the process to others and to reproduce the same judgement consistently (Brehmer, 1986). It is also curious to note that, even though people do not often understand their own quasi-rational inference process and think it is complex, it can often be reproduced by simple mathematical models. Such models have revealed that different people choose very few but different cues and put different weights on common cues. But they are likely to combine them in a way that can be replicated using a simple additive model (Hammond *et al.*, 1975; Cooksey, 1995). However, talk of discrete cues and aggregating models must not mislead us. The perception of the cues is holistic—all the cues are processed together. This is why most

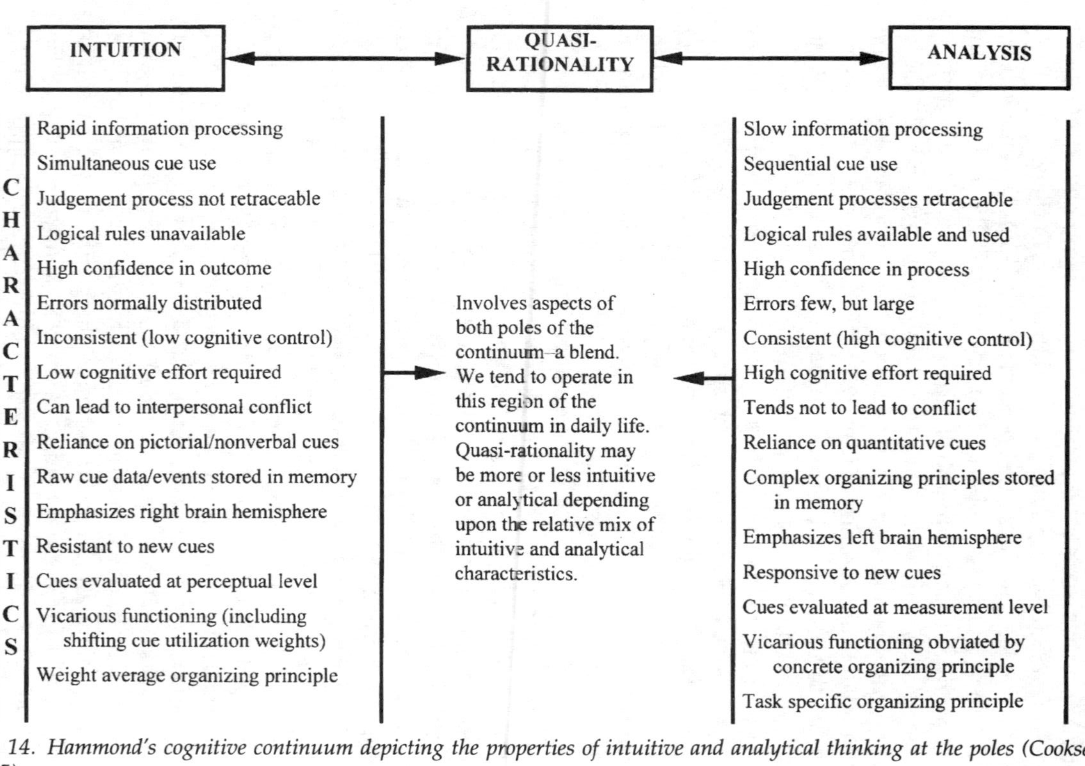

Fig. 14. Hammond's cognitive continuum depicting the properties of intuitive and analytical thinking at the poles (Cooksey, 1995)

people choose only a few cues out of a large possible array. The brain has limited capacity to integrate more than (perhaps) half a dozen factors at once. Under these conditions the use of intuition is important.

It is perhaps wise not to define intuition too strictly because the word may cover all those skills of perception not recognized as analytical. Nevertheless, experts have developed a form of intuition which is very useful in judgement. When a professional becomes expert at a task, he or she develops an ability to recognize patterns of cues such that the interface process can be greatly shortened (Simon, 1986). Such an ability is used with great effect in circumstances where a rapid response is required and is described in some detail in chapter 8.

However, the question of our ability to make better judgements with experience is not straightforward. Feedback is required, otherwise we cannot adjust our judgements over time (Brehmer, 1986). This explains why the judgement of older people in jobs where situations are repeated often, and the results are readily observed, is often better than that of younger people. We should expect the judgement of a fireman with ten years experience to be better than that of a new recruit. Unfortunately, managers are often required to make judgements in new circumstances and feedback is often minimal. This implies that, as your experience as a manager increases, your judgement should improve in more familiar situations but remain at its innate level in unfamiliar situations. There is some evidence for this. Kline (1994) tested a wide range of managers of varying experience using case studies. He found that managers with higher levels of education and experience were not necessarily better at making effective judgements in strategic decision making simulations.

Our cognitive abilities to make sound judgements are influenced by perceptual rules of thumb, or heuristics, which appear to have been developed to enable us to simplify the deluge of information pouring through our sense organs. An example given by Tversky and Kahneman (1982) is our judgement that the less clear an object appears the further away it is. This makes sense as a general rule but can lead to errors in particular situations—such as driving in foggy conditions. Many of the errors and biases in our statistical judgement, noted in chapter 4, appear to be produced by these rules of thumb, and others will be discussed in following sections. Our conscious or unconscious decision about how to make a judgement in any particular circumstances is, as noted previously, influenced by the nature and complexity of the task. It is also influenced by a trade-off between the need, or otherwise, for accuracy and our perception of the effort required to achieve it (Payne *et al.*, 1993). For example, the use of a rule of thumb that ignores much of the relevant information may be used, even though we know the result may be flawed, because time or weariness does not allow us the luxury of a more considered approach. However, this research also indicates that we are highly adaptive and opportunistic in our response to decision situations. We appear to have a hierarchy of judgement strategies which we apply (generally successfully) in an intelligent manner to suit the changing circumstances.

Decision and action

It is possible, of course, to make a judgement without making a decision. We are constantly judging the world around us but only rarely do our judgements require action. Similarly, we make decisions against 'our better judgement'—often motivated by an emotional response to the situation. Translated, I supposed this to mean that we have been faced with a new cue which has triggered a new and different judgement. The speed of this response to new data may give us the impression that the previous careful thought was wasted and a certain amount of guilt may result. Although this may be a common fact of personal life I am assuming that the effective manager does not often succumb to spontaneous and arbitrary decision making!

Decisions are sometimes not followed by immediate action. In which case the action becomes latent—a plan. Rather more common is the integration of decision making and action in dynamic situations. For example, the production of sketch designs on a drawing board involves a complex interplay of cue generation, judgement, decision and action such that words, the eye and the hand can give the impression that all the processes are occurring simultaneously, or sometimes, in the wrong order. Indeed, in most real decision making the process outlined in Fig. 12 is iterated many times before the process is concluded, with loops back from any point in the cycle. Figure 13 shows just such a loop in the thinking and judging stage. Having made a judgement about the problematic situation, it is henceforth impossible to reassess the situation without it being affected by the modelling inherent in the process. The situation has become the cues and we will find it difficult to see the situation anew (Brunswik, 1952).

Some of the many influences on the process of making personal decisions will be discussed briefly in the following sections, with an emphasis on what has been learnt from empirical research.

Influences on cue selection and use

We have noted before that the human animal has developed some efficient strategies to avoid the limitations of the brain. However, we have also developed a number of ways of thinking that can lead to biases in our decision making. Some of these psychological effects will now be discussed in relation to the selection of cues for judgement. For this I will rely partly on Hogarth's (1987) summary of the experimental data available on this issue. Biases in our selection of cues include:

- *Availability.* We recall with some ease events seen on television or anecdotes told at a party. These highly retrievable pieces of data may cause us to put too much weight on associated cues.
- *Selective perception.* Your role will to a large extent determine what cues you perceive. Engineers will tend to see technical cues and politicians will see social cues.

We may also disregard information that does not fit our preconceptions and give too much weight to confirmatory cues.

- *Concrete information.* We are more likely to value information derived from our past experience, or those of trusted colleagues, than the abstract information presented in a report.
- *Illusory correlation.* We sometimes jump to the wrong conclusions about the value and function of a cue because we mistake correlation with cause. A classic example would be the attribution of an improvement in performance of an employee to your efforts, when all that is happening is that the dip in performance has bounced back to above the mean due to natural random effects.
- *Data presentation.* We remember information presented first or last in a list. In accordance with our professional backgrounds we may favour diagrammatic presentations over written descriptions, and the beauty and logic of a data display may blind us to its inadequacies. This latter problem is becoming more acute as multimedia presentations become more common.
- *Wishful thinking.* Our personal preferences may inflate the importance of a cue beyond its real significance.
- *The halo effect.* This is the effect of one particular cue on another—one casts a halo over the others. We are used to experiencing cues in clusters such that one thing goes with another, so that we tend to reject or rationalize any cue that appears to contradict the consistent set.

Not mentioned in the psychological literature is the effect on cue selection of our value differences and the norms of our societies. The importance of such rule-following to an understanding of decision making cannot be overemphasized, as rules are without doubt, the single greatest shaping mechanism in the management decision process (March, 1994). Three sorts of rules will be briefly discussed.

1. *Personal values* are integrated into our personalities during childhood and later socialization (Trainer, 1982). They are derived from the enacted values of our parents, friends, school teachers and colleagues, and are very long lasting. These values, mixed with other personality variables and the other secondary socializations of adulthood, determine our personal attitudes towards problematic situations. The result is that we look for moral cues that tell us whether a course of action will or will not violate our value system. Is this a situation that I, in all conscience, can work within? For example, cues that indicate that corruption is about to take place would cause many to opt out of decision role in that organization. They would decide to withdraw.
2. *Organizational norms.* These are not as stable or long lasting as personal values but, nonetheless, are very important to the decision maker. Cues are sought that indicate how a decision should be made given the rules, regulations and culture of your organization. Is this the right way for a professional in my role to behave? Sometimes organizational norms and the preferences of the individual may conflict and, provided some fundamental personal value is not violated, most conscientious managers will do what is deemed best for the organization (March, 1994).
3. *Cultural norms.* In our modern world we are all aware that national cultural characteristics effect styles of decision making. What is less obvious is the effect cultural norms may have on cue selection and the weight placed on these cues. At the simplest level, road realignment decisions may be required because of religious or taboo cues, or meetings postponed because of unlucky days or a national race meeting. More subtle differences may arise due to characteristics such as the individualism or collectivism noted by Hofstede (1984). Thus, an individualistic Australian, negotiating with a collectivist Japanese group, may *read* behaviour in the meeting as indicating agreement when there is none, rudeness when none was intended, and (the worst mistake of

all) silence for agreement. He or she may assume (see) that the negotiation has reached a stage that requires a decision. The Japanese may interpret the same cues as meaning that both sides must withdraw and consult with their constituencies. The potential for mistakes are numerous—the same problematic situation may present quite different cues, similar cues may have different meanings and the assumed decision process may, in practice, be quite dissimilar.

Influences on quasi-rational judgement

At the risk of completely destroying your faith in your own judgement we must continue with our review of the way that an individual cognition can bias judgement. Having looked at the difficulties of cue selection we will discuss some other factors that affect our quasi-rational judgement (Hogarth, 1987; Mullen and Roth, 1991).

Framing effects

If we are told that we can save $5 on a $15 purchase by driving to another branch that is holding a sale, we are more likely to do so than if we were promised a saving of $5 on a $150 purchase. Yet the effort expended on driving is the same. Our choice may be different because the $5 is presented in a different frame of reference. Tversky and Kahneman (1985) present another demonstration of the framing effect. One group was asked to consider the following decision.

> Imagine that you have decided to see a play where admission is $10 per ticket. As you enter the theatre you discover that you have lost a $10 bill.
>
> Would you still pay $10 to see the play?

88% said they would see the play.

They asked a second group of subjects

> Imagine that you have decided to see a play and paid the admission price of $10 per ticket. As you enter the theatre, you discover that you have lost the ticket. The seat was not marked and ticket cannot be recovered.
>
> Would you still pay $10 for another ticket?

Only 46% of the second group would have bought another ticket. The majority of this group apparently framed the problem as a doubling of the entrance cost whereas the first group framed the $10 as a marginal loss of capital. In the same way we may pay several hundred dollars more on extras when purchasing a car, but would be reluctant to spend similar amounts for fittings in our existing car. And we may even query a lunch bill of $20 after spending $20 000 on the car and extras. The utility of money (or anything else) seems to depend on the mental account we place it in.

Overconfidence

Generally, people have a displaced confidence in their judgement. When asked general knowledge or probability questions, experimental subjects performed worse than they thought they had (Slovic *et al.*, 1982). Calibration experiments that test the match between confidence and accuracy of judgement, demonstrate that those without training and feedback perform badly. Lichtenstein *et al.* (1982) found that from 15 000 judgements, when subjects were 98% sure that an interval contained the right answer they were wrong 32% of the time. Even experts are prone to some overconfidence. Hynes and Vanmarke (1976) asked seven geotechnical gurus to estimate the height of a trial embankment (and their 50% confidence limits), that would cause a slip failure in the clay bed. Two overestimated the height and five underestimated. None of them got it within their 50% confidence limits. The point estimates were not grossly wrong but all the experts underestimated the potential for error.

Irrelevant learning
We gain a false confidence in our judgement when the outcome appears to be positive. This may be due to chance or we may have rejected alternatives that would have produced a better outcome. This inaccurate learning then affects our subsequent choice of cues. We may also attribute a successful outcome to good judgement and an unsuccessful outcome to bad luck. Moreover, our bad memory may produce a quite logical but false reconstruction of events. Finally, hindsight makes the past seem so inevitable that we forget all the other alternative paths events could have taken.

Cognitive dissonance
This is that uncomfortable feeling we experience when information conflicts with a deeply held value or belief. This distress may cause us to dismiss the truth or rationalize the situation by rethinking our beliefs. We are looking for an easy way to eliminate the contradiction. This, of course, may well lead to a biased judgement of the cues. After a poor decision outcome is known we may go through the curious ritual of bolstering. This is an attempt to reconcile our favourable self image with the negative results of our decision. We quickly bolster the positive aspects of the decision and talk down the importance of the poor consequences.

Sunk costs
If a decision making process is going wrong and considerable time and energy has gone into the process so far, we are tempted to finish. This of course, contradicts the principle that the costs of reaching the present moment are gone (sunk) whatever we do, so it is wise to judge the utility of continuing only on our assessment of the current situation. Our judgement is often badly affected by regret, and if the only reason we are continuing is to justify our loss, then the logical procedure is to stop, forget the past, and start again using a new method.

Stress
When we are under stress we find it difficult to concentrate fully on the decision process. Our search for cues is often prematurely

curtailed and our inference abilities blocked. We are either worried about something else or worried about the decision we are about to make. In fact, decision making itself can be stressful and the consequences of this are more fully explored in the next chapter. Under stress we have to rely on our learned capabilities to carry us through. Expertise is therefore important to good decision making—a subject again discussed in chapter 8.

Influence from others

These may be good or bad in their effect. Conformity is rarely a good thing when trying to recognize the nature of a problematic situation, because the potential cue array tends to be restricted. Alternatively, conferring with colleagues can often widen the range of cues considered. Whether group decision making is more effective than decision making by individuals is an open question, much dependent on particular circumstances. Chapter 9 will contain a full review of these issues.

Personality

Jaccard *et al.* (1989) listed the measurable individual personality characteristics that contribute to efficient personal decision making ability. The results are somewhat predictable. Personality characteristics thought to help an ability to make good decisions are listed below followed by less useful traits.

Helpful characteristics include

- an internal locus of control—the feeling that you, rather than fate, controls your destiny
- endurance
- flexibility
- responsibility
- risk taking
- a sense of order
- human understanding
- innovation
- a sense of achievement

- self acceptance
- critical thinking
- complexity of thinking
- intellectual efficiency.

Ineffective decision makers are often weak in these areas and share some negative characteristics, such as impulsiveness or procrastination, conformity, orthodoxy, rigidity, and anxiety or depression.

Such a list is of little help if we are attempting to predict behaviour, for it is the interaction of these characteristics that will produce a distinct decision making personality. In an attempt to overcome this difficulty some researchers have tested portfolios of characteristics to define different decision styles. A good example of this genre is Rowe's decision style inventory (Rowe and Boulgarides, 1992). A set of 20 questions are asked of the subject about such things as personal objectives, expectations of others, communication methods, analytical or intuitive task representation, response to situations, response to other people, attitudes to work, response to stress and response to control. Depending on the number of ticks in each of four columns the subject's decision style is rated as more or less analytical, conceptual, directive or behavioural. The decision maker can then compare his or her scores on these decision styles with those of samples of male and female managers, chief executives, admirals, foreign executives and students.

However, despite the attraction of such an approach for those in search of self knowledge, Goodwin and Wright (1991) have confidently stated that 'the overall result of attempts to delineate distinct cognitive styles of decision making that characterize individuals over a variety of decision situations or tasks has been, to date, disappointing' (p. 206). The reason for this is that it is the interaction between the characteristics of the decision maker and the characteristics of the task that determines good or bad outcomes. Such interactions are potentially large in number. Small changes in a task can produce evidence of good or bad decision making so that, overall, 'decision making and judgment seem contingent on the nature of the decision making situation' (Goodwin and Wright, 1991, p. 209). Nevertheless, a check on decision styles may be very useful when attempting to match people to jobs that very obviously require a particular type of personality. It is clear, however, that flexibility of approach is important to match tasks with styles, and those rare individuals that can span a number of styles have an advantage in the world of dynamic managerial decision making (Nutt, 1993a).

Summary

- The decision process may be modelled as enframed by organizational goals and progressing from a recognition of a problem or opportunity to a quasi-rational thinking process which results in a judgement. Decision and action are the active responses to that judgement.
- Judgement requires the perception of situational and problem specific cues which are selected, weighed and processed in a different manner by each of us. Cue selection and use is heavily influenced by a range of psychological factors, personal values and social norms.

8

Individual decision behaviour

Reasoned choice

Despite what we have learnt from the previous chapter it is still commonly assumed that individual decision making, particularly in organizations, is associated with choice made after a process of fully rational reasoning (Zey, 1992; Janis and Mann, 1977). A typical reasoned choice procedure would be as follows (Jaccard *et al.*, 1989; p. 100)

- *Problem recognition.* The individual determines that a problem state exists and that a decision must be considered.
- *Goal identification.* The individual specifies *a priori* the purpose of the decision: that is the ideal outcome of the decision.
- *Option generation/identification.* The individual thinks of potential alternative solutions to the problem at hand.
- *Information search.* The individual seeks information, either about what additional options might be available or about properties of one or more of the options under consideration.
- *Assessment of option information.* The individual consciously considers the information they have about the

different options. Based on this information, the individual forms preferences for some options relative to others.
- *Choice.* The individual selects one of the decision options for purposes of future behavioral enactment.
- *Post-decision evaluation.* The individual reflects on the decision after the option has been enacted, then evaluates the decision (and the decision process) in light of the outcomes that have resulted.

As James March (1982) has pointed out, this process optimistically assumes that the decision maker has an unambiguous knowledge of alternatives; a knowledge of consequences; a consistent preference ordering and a decision rule. Moreover, in much of the decision theory literature it is also assumed that the procedure and decision rules are directed towards some form of optimal outcome in relation to the goals. The fact that the day-to-day experience of ordinary decision makers does not conform to such an optimizing model is well-known. In practice, we may admire its austere beauty but feel inadequate before its demanding complexity. What actually happens in real-world decision making situations often produces suboptional outcomes and will be the subject of this chapter.

Constraints on reasoned choice

We have learnt from the previous chapter that choice is not always part of a decision process. Moreover, psychological and behavioural research in organizations has indicated that reasoned choice is constrained and modified by a number of factors.

Complexity is confusing

It was in the late 1950s that Herbert Simon (1976) first articulated the notion that human beings were not very smart. Unaided, we are constrained in our capacity to fully perform the reasoned choice procedure by our cognitive limitations—we cannot assimilate the volume of information required or do the mental computations needed to come to an optimal conclusion. Therefore we simplify. If we do not achieve a goal we search for information just long enough to satisfy that goal—in the words of Simon, we 'satisfice' rather than optimize. New alternatives are sought in the area of old solutions and search is discontinued when an improvement is made that just satisfies some modest change of goals.

At its most dangerous, satisficing takes place using very simple decision rules—like accepting the word of only one doctor when making important personal medical decisions despite the possible gains to be achieved by obtaining multiple opinions. But as Janis and Mann (1977) go on to point out, a single decision rule may not always be used. Elimination-by-aspects is a common satisficing strategy that uses a combination of simple rules (Tversky, 1972). For example, one may choose to buy a house by eliminating all above a certain price, then those without three bedrooms, then those without a sizeable yard, and so on until the choice is narrowed to a point where the remaining aspects to be compared are within the reach of our bounded rationality.

Lindblom (1959) has described the form of satisficing that takes in large public bureaucracies, where there is great reluctance to radically change the value systems embedded in existing policies. The response is to use a strategy of 'successive limited comparisons', which later became known as 'incrementalism'. No grand theoretical notions are used and explicit goals are rarely stated. Reasoned choice is not performed. Existing policies are valued because of the social acceptance of the value system embedded in them. Thus, new policies are chosen that only alter the value systems by a small amount and attention is concentrated on the differences between the old and new policies. Undue conservatism is avoided by the influence of competing value sets espoused by other interests in plural democratic systems. This process is less than smooth, as attention and energy fluctuate in the decision environment. Twenty years later Lindblom (1979) indicated that this 'disjointed incrementalism' was but one important approach to practical decision making that varied on a continuum from simple satisficing to what he called 'strategic analysis'. Even at the strategic analysis end of the continuum, comprehensiveness is not attempted but rather a limited number of possible solutions are selected for careful analysis. Lindblom (1959) admits that incrementalism implies that important possible outcomes may be neglected, alternative possible policies will remain unexplored, and other affected values ignored. Nevertheless, it may be the best we flawed humans can do in complex and ambiguous administrative situations.

There are other reasons (perhaps related to our cognitive limitations) for our indifference to the reasoned choice mode.

Information may not be critical

The reasoned choice model emphasizes information search to provide the raw material for decision making. But the observation of real organizations indicates that

1. information gathered is not used
2. decisions are often made first and information sought to justify the solution
3. much information gathered is irrelevant to decision making.

March and Shapira (1992) explain that information is gathered in organizations for many reasons other than to aid decisions. In particular, it forms a medium for general intelligence concerning the organization's functions, and serves as a symbolic legitimator of management competence. Moreover, there is also an element of corporate wisdom at work which recognizes that information is often wrong or is being used strategically to advance personal or group interests.

First choices are rationalized

Although it is not explicitly stated in our model of reasoned choice, it is normally assumed that the actual choice between alternatives is achieved using some rationally defensible method. However, if a computational strategy is not used, the process is often flawed. Even if a person is given all the information required for choice and the time and environment for a calm appraisal, the method adopted is often biased by the process of rationalization.

Montgomery (1993) has observed in laboratory and real-world situations that decision makers faced with a choice between alternatives, tend to look to dominance for guidance. The first step is to reduce the number of alternatives by a process of elimination by aspects (described previously). The person then picks a promising alternative because it looks the best on one important aspect. If this alternative proves to be equal to all others in most other aspects, then it is chosen. It is dominant. If it is not obviously dominant, the decision maker tends to reinterpret its standing in relation to the others by de-emphasizing some aspects, enhancing others, trading off some good and bad aspects and finally, clumping some aspects under a more comprehensive heading. If by this process the decision maker can be convinced that the favourite is really dominant, the stage is set for action. Thus, the premature commitment to an alternative may quickly subvert the reasoned choice procedure and effectively waste the considerable effort expended on information search.

Goals may be ambiguous or in conflict

It is hard to predict the future and sometimes such predictions may be counterproductive. Being too prescriptive about future preferences may stand in the way of the flexibility of response that is necessary in a rapidly changing world. Moreover, being less than precise about organizational objectives may allow a more fruitful and open interplay of conflicting preferences. Thus, as many management consultants have discovered, strategic planning is often honoured more in theory than in practice.

Conflict between goals is rarely settled by a process of weights and trade-offs. Decision making is part of organizing, and organizing is a political game. Power will be exercised, manoeuvres will occur, and interests will be defended by whatever means are necessary (March and Shapira, 1992). Indeed, whether a problem is recognized as worthy of inclusion in the decision process and, if so, in what form, is one of power's many manifestations (Lukes, 1974).

As Cyert and March (1963) have observed, conflicting goals are usually not addressed all at once. One goal is addressed at a time and the incompatibility of goals at some higher level is ignored. This requires a search for only that information that has a bearing on the local goals, and often results in the acceptance of only data produced by search in the immediate problem vicinity. This process the authors called the 'quasi-resolution of conflict' and is the recognizable attribute of most large organizations. Luckily we have the wit to live with less than perfection.

Decision making may be experimental

The rigid rationality of the reasoned choice model does not cater for the experimentation and social learning that goes on in organizations. In good times, in particular, managers are encouraged to take risks, try new things, make changes and generally to exploit the ability of the organization to absorb mistakes. In this way surprising new strategies may emerge that could be useful in the bad times. It is a long term investment in foolish decision making (March, 1982).

Decisions may be random

It is assumed by advocates of the reasoned choice methodology that goals can be readily identified. This may be so in most circumstances, but sometimes decisions are made in an environment of conflicting subgoals and unclear organizational processes such that the method of arriving at a solution cannot follow the conventional means–end procedure. In these ambiguous circumstances decisions may be a function of a random coming together of problems and solutions in meetings or other such 'garbage cans' (Cohen *et al.*, 1972). People wander in and out of decisions, as the complexity of organizing demands their attention, contributing whatever energy is available to any particular problematic situation as they present themselves to them. They may carry with them a portfolio of solutions looking for problems and problems looking for answers. As participants come together in choice opportunities so available solutions may come together with suitable problems. The key to decision making in such organizations is to maximize the number of meetings rather than attempt a process of rational modelling.

Decisions may be symbolic
How do we explain that

1. as noted before, information is ignored
2. people are hungry to be asked to take part in decisions but then do not do so
3. people spend very little time actually making decisions
4. people will fight hard for a policy and then be indifferent to its implementation?

In response, March and Shapira (1992) contend that much decision making is not about arriving at decisions, but rather has a symbolic role in organizational life. Thus, a decision is an opportunity for

- exercising organizational procedures and role duties
- defining organizational meaning
- distributing rewards or blame and the general exercise of power and status
- socialization, bonding and the enjoyment of being a decision maker.

In this manner, a common sense of meaning and purpose may be the true result of a decision making process rather than a more overt utilitarian outcome. The same processes may be also a major arena for the demonstration and resolution of power. Above all, decision making reinforces the legitimacy of management—managers must be seen to be decision makers. This often involves going through the decision making motions after action has already been taken. We should be careful, therefore, not to inflate the importance of decision making in the performance of organizing. Sometimes, in the words of March and Shapira (1992), 'trying to understand decision making as a way of making decisions may be analogous to trying to understand a religious ceremony as a way of communicating with a deity. Both characterizations are correct but both are misleading' (p. 290).

Decisions make cowards of us all
Judgements can be kept to oneself. Decisions, on the other hand, are a commitment to action—action that can be seen as beneficial or disastrous, and which may be traced back to us. We will be judged, in part, by our decisions, and our careers may depend on a reputation as a sound decision maker. The stakes in decision making are high, and courage is required when we are required to face their consequences. Decision making is, in the words of Janis and Mann (1977), a 'hot cognitive process' involving heightened adrenalin flow and apprehensiveness. This tension is exacerbated by the time and social pressures under which we have to make organizational decisions. The two researchers found that only under conditions of little pressure would a process akin to reasoned choice take place. At other times attempts will be made to choose solutions that diminish the potential threat to oneself or, perhaps, to devise some means of avoiding the decision. If the decision is important and the time and resources short, the decision maker may go into a state of sustained panic and highly flawed decisions are to be expected.

Janis (1989) also describes the way that decision making is constrained by the fear of offending powerful members of the organization or of breaking organizational norms. As discovered by Jackall (1988), managers often work in an atmosphere of fear and in these circumstances it is understandable that they may wish to avoid blame by going along with the boss.

The decision maker has obligations
In most day-to-day decision making in organizations the dominant decision process is unlikely to directly involve the interests of those involved. Decision makers are people in roles—doing jobs that are described and constrained by rules. Indeed, sometimes they may be required to act against their own best interests for the good of the organization or the public. Their model of decision making is

not one of reasoned choice but rather one of obligatory action. Instead of setting goals, deriving options and making choices free of constraint, they consciously or unconsciously perform a role. Thus, the decision maker may approach the problem in the following way (March, 1982, p. 35)

- What kind of situation is this?
- What kind of person am I?
- What is appropriate for me in a situation like this?
- Do it.

The fact that an individual may act in ways that apparently conflict with aspects of their personal value systems is not surprising or unusual. We all carry around a mixed bag of values associated with our various life roles, some of which will be in conflict. Luckily, when we get to work we forget our role as loving father or football fan and become submerged in our job. Our professional training, socialization and acculturation is usually sufficient to ensure that we will use rules appropriate to our job (March, 1994). Indeed, the process of organizing would be very difficult without the confidence that staff members will act in accordance with their obligations.

We will now look at models which, to some degree, take account of our objections to the reasoned choice model.

Two alternative models

Image theory

Image theory (Beach, 1990; 1993) is a rule-based decision model that includes the possibility of choice. It does not directly address the question of problem definition or, indeed, information search or alternative generation. Rather, it describes a mechanism for sieving potential solutions and progressing plans for implementation. The lack of attention to the earlier aspects of the decision process is no doubt due to its laboratory origins. Nevertheless, 15 years of experimental work and extensive verification in real-life settings, involving both personal and organizational decisions, indicates that this model must be taken seriously.

The decision maker approaches the decision scene equipped with a *frame* consisting of a set of schematics (knowledge structures), about such situations. The situation is recognized and the appropriate frame is put in place. This frame defines the *status quo* and any decision making will result in a change to the frame. As the *status quo* is highly valued, decision making is undertaken with great reluctance.

The schemata within the frame may be divided into three images. The *value image* contains all the personal and organizational values and norms relevant to the situation. In an organizational setting, these include role values and corporate rules and obligations. The *trajectory image* contains the goals to be achieved and the *strategic image* contains a portfolio of plans, forecasts and methodologies for the implementation of any solution.

Potential solutions, in the form of new goals and plans, are considered sequentially and tested for compatibility with the existing images. Thus, solutions close to the *status quo* will be readily adopted. If the fit is not perfect the goal or plan may be changed. If the solution is intolerably incompatible it will be rejected and another tested. Occasionally, more than one solution will pass the compatibility test and a choice procedure of some sort will take place. It is noted by Beach (1993) that the greater the complexity of the choice, the more likely it is that decision makers will use non-compensatory intuitive methods. In most cases of choice they rely primarily on dominance as a criteria.

In a similar manner the feasibility of implementation is tested against the three images. In particular, the question of whether the derived goal will, in practice, be achieved involves the testing of imagined stories or scenarios of the potential progress of the plan.

Image theory is a general model that explains decision behaviour in most situations. The following model concentrates on the expert under pressure—a situation familiar to most engineering managers.

Recognition-primed decision theory

This model (Klein, 1993) is the result of the study of the protocols and behaviour of fire chiefs, tank commanders and, significantly for ourselves, design engineers. This form of decision behaviour takes place in settings familiar to the decision maker and in circumstances where situational expertise can be mobilized. It often enables rapid responses to critical situations. In these circumstances the reasoned choice rationality is not used—in fact choice does not feature at all. Judgement is the key.

The first stage of the decision process consists of *situation recognition*. This requires the recognition of cues, typical to the decision maker's domain, that describe the type of situation, its possible causes, and its likely development. Based on these cues, the decision maker sets achievable and appropriate goals (potential solutions). These action alternatives are taken one at a time and not compared with each other. Rather, the most typical option is first selected and a mental simulation of the likely results is used to evaluate its suitability. If the solution is deemed adequate, it is adopted immediately. This action may be modified as the situation develops and radical revisions will stimulate more imaginative simulation of likely outcomes. Thus potential problems are rehearsed and likely objections addressed ahead of time.

This model indicates the importance of domain-specific knowledge for efficient decision making in normally stressful situations because, by the time novices have used some reasoned choice method, the situation has often changed and the decision opportunity is lost. However, where the decision maker is faced with an unfamiliar situation the use of more analytical methods involving choice may be more appropriate. Nevertheless, recognition-primed decision making may be very suitable for engineering managers in an engineering setting.

Caution

It should be emphasized at this point that, despite the number of circumstances in which reasoned choice may be inappropriate or frustrated, the approach is both powerful and useful. Where enough time and resources are available, the data is abstract and can be combined, a number of interests must be satisfied and, above all, where the decision must be publicly justified, then reasoned choice is an efficient tool.

Managing your decision making

Up until this point we have concentrated on a description of what is known about decision making behaviour. Now it is necessary for me to stick my neck out and be just a little prescriptive on the subject. I promise, however, that I will remain close to what can be deduced from the preceding discussion.

As we are talking of individual decision making in an organizational setting, I will use our chapter 7 model of the decision process as a framework. Stripped of its influences this describes five interrelated stages

1. problem or opportunity recognition
2. thought (search and quasi-rational inference)
3. judgement (end result of inference)
4. decision (intended action)
5. action.

You will note the generality of this model, in that judgement may or may not involve alternatives—the judgement may be of the merit of only one possible solution in relation to the circumstances.

In our review of what is required to make effective individual decisions, we must make some common sense assumptions. The first assumption is that engineering managers will normally be making decisions in circumstances where their expertise can be mobilized in the process, but occasionally they will be required to make decisions where expertise in engineering management will give little guidance. Secondly, that they can judge whether they will have the time, resources and motivation to go through an approximation of the reasoned choice process. This is not always obvious in the busy life of a manager. Lastly, we will assume that decisions are not of a long term strategic nature or required to be made by a group. These circumstances will be covered in subsequent chapters.

Circumstance I.
Decisions in your professional domain

First response
Always make a recognition-primed decision first.

- *Problem recognition.* You will recognize the problem as one within your domain of expertise.
- *Thought.* Think of the possible causes and the likely problem development.
- *Judgement.* Use your experience to choose a typical, practical, solution option.
- *Think again.* Run through the likely consequences of adopting this solution to look for drawbacks. If the drawbacks look significant repeat the *judgement* stage.
- *Decision.* If everything is reasonably satisfactory with your mental simulation, adopt this solution.
- *Action.* Keep observing the situation for new or changed cues that will require a change of plan. If new information makes your chosen solution untenable, abandon it and treat the situation as a new decision opportunity.

Remember, this is a *satisficing* solution—it is designed to find a solution that can be seen to be reasonable. If you feel you are required to demonstrate that you have attempted to find the *optimal* solution, and you have the time and the resources, put this recognition-primed solution aside after the decision stage. You can always implement it at short notice if you run out of time.

Second response
If you require an optimal decision it is probably wise to use a combination of the reasoned choice and image theory models.

- *Problem recognition.* Use your imagination to think through how the other interested actors may view the situation. You may have neglected some social factor (for example) that could be vital to the nature of possible solutions.
- *Thought.* Think carefully about the goals you are hoping to achieve as these will serve to structure the problem definition and hence the range of alternative solutions. Gather as much data as possible about the technical, organizational, social and personal factors that are likely to influence choice. Cluster the information into not more than five categories (cues) for ease of judgement.
- *Judgement.* Compare options, as described by the data clusters, to the problem-related values, goals and plans of your organization and society. This is a holistic, judgemental process heavily dependant on the mobilization of your expertise in this domain. Write it down for the record.

  If two solution options appear to be about equal look for a characteristic that would produce dominance. In particular, play around with the cues a little to test the importance of each to the solution. Do not be fooled by rationalization at this time. Above all, make sure that the chosen solution does not violate any organizational norms or it may be difficult to implement.
- *Action.* Implement and monitor.

You will note that I have not suggested the use of a technique, such as Decision Analysis, which requires a comprehensive disaggreggation of the data, an axiomatic decision rule and a reaggregation around a single utility unit. Although this is possible, it is probably unnecessary in these circumstances. We are operating within your domain of expertise, which allows, provided the number of cues is limited, the efficient use of holistic judgement. This strategy may not be as (apparently) precise as numerical methods but it is not as prone to producing disastrous mistakes. Why waste your expertise and your time? Of course, if you have personal expertise in these analytical methods and you have time, and the complexity of the problem justifies the effort, then by all means use the result as another important cue for judgment.

Circumstance II.
Decisions outside your professional domain

Do not do it alone—get help. Treat this as a circumstance that will require a group decision as described in the next chapter.

Individual decision management—an illustrative case study.
'The Falsework'

Don liked being a resident engineer on big civil jobs. They lasted long enough for his family to settle down a little—and they didn't involve architects. But today he wished he was back in the public service with a nice comfortable set of rules to follow. What was bothering him was not unusual—inspectors needed backing up on occasions. But, this particular day the chief inspector had asked for an opinion—which probably meant that the inspector felt that the situation was going to generate considerable strife. And he was right. Whatever Don decided the outcome could be nasty.

The inspector had taken Don to see the final piece of arch falsework to the access tunnel. This section of work was critical to the progress of another contract which involved a high level cross passage over the final section of arch. Without a pour this week a large claim was probable and 'his' contractor was working flat out to make the programmed completion date. The pours were heavy but normally the falsework was easily assembled on the invert arch and few problems had emerged in the last four months of work. Today, however, one corner of the falsework was not firmly supported on the invert and as many as one quarter of the scaffold foot plates were supported by the newly excavated soil very close to the edge of a two metre drop off. Although sand bags were being used to stabilize the face, the whole thing looked unprofessional. This sort of cantilevering was not unusual in skewed arch situations but Don was nervous because the stakes were high. If he stopped the job the contractor would go berserk and no doubt a claim would be generated which would involve the client. If he let it go and the falsework failed he would be in even deeper trouble.

The fact that Don had seen this situation before led him to conclude that the falsework would probably be OK. He was inclined, therefore, to let the pour proceed. But experience also warned him that the easy way was perhaps too easy to take. Under these circumstances he had a little trick that had served him well in the past. He imagined what the chairman of his consulting company would expect him to do. What procedures would he expect of one of his top engineers? What things should he consider? Would he be able to look the chairman in the eye? How would the future of the firm be affected? Next he imagined himself in court defending his actions in front of a judge. What action would the judge consider endangered life? What decision could be considered negligent in the circumstances? All in all, the scenarios indicated that some cautious action was required and Don braced himself to call the contractor's site manager. Maybe they could sort out a solution that wouldn't delay the project too much.

Discussion

It is clear that this is a decision well within Don's area of expertise. Nevertheless, the non engineering consequences for himself are causing him some trepidation. His initial reaction—his recognition-primed decision—was based almost entirely on technical cues and he concluded that the probability of failure was small. However, he was wise enough to pause and think through whether his initial reaction was the best response possible. At this stage he had moved into a decision making mode something like that described in image theory. He was using his imagination to look at the situation through the eyes of others he respected. The cues that dominated his new perception were those associated with his social identity. These were strong enough to change his judgement and set him back on the track defined by his role as a resident engineer.

Summary

- Despite what we have learnt from the previous chapter it is still commonly assumed that decision making involves a fully rational reasoned choice process which, based on comprehensive information, seeks to optimize an outcome.
- Behavioural research has indicated that managers rarely follow this procedure for reasons that reflect the ambiguous and complex relationship between individuals and organizations.
- Empirical research has indicated that image theory and recognition-primed decision theory may be better guides to real individual decision making.
- The chapter concludes with decision management recommendations based on the models discussed previously.

9

Group decisions

For the purposes of this chapter, I will define a group as two or more people brought together for some organizational purpose. A team is a group that has worked together for a time sufficient to develop some degree of unitary behaviour. *Ad hoc* associations such as meetings cannot, under our definition, be considered a team unless they have worked together long enough to coordinate tasks. Thus, eleven young people playing soccer together becomes a team when they have practiced sufficiently to complete manoeuvres together without complex communication. They are aware of desirable things to do in relation to others in any situation and can anticipate the behaviour of other team members. The other characteristic of a team is the interdependence and coordination of members with different interests, knowledge-bases and roles. Thus, a group may be made up of interchangeable people—a jury for example—but a team welds together different roles. Finally, groups may be brought together specifically to make participatory decisions but teams only make decisions as part of a more global function.

Myths and reality in group work

It is part of management folklore that participation in group work is good for the well being of an organization. In particular, job satisfaction is increased if participation is encouraged. Such beliefs persist despite ample evidence that group work is stressful for many who are forced to take part. Moreover, little evidence exists that performance is improved when work becomes more enjoyable. It is also possible that many find groups a convenient way to avoid work (Sinclair, 1992). Nevertheless, the corporate culture fad sparked by the success of Japanese business has entrenched the view that groups are good things and, as a consequence, we must learn to understand their dynamics. Certainly, for engineers, group decision making in project management teams is an important issue.

Despite the hype, some things we can say with some certainty (Mitchell *et al.*, 1988)

- group work is slow
- groups contribute a wider range of cues for judgement
- participation is likely to create increased commitment to a decision
- the symbolic and political returns from participation may be high.

Other aspects of group performance are unclear, but we do know that the role of leadership is of critical importance. As Sinclair (1992, p. 618) notes, 'recent research confirms that the most critical ingredient of team success is its leadership ... the abdication of leadership can, in effect, paralyze groups'. Clearly, however, the role of leadership in groups and teams is different. In an *ad hoc* meeting, leadership may consist of canny chairmanship or raw power. In the transformation period between group formation and the achievement of teamwork, leadership may have the function of aligning individual interests and interpretations—providing, in fact, a vision or model of behaviour. When the team achieves a satisfactory degree of integration and unitary purpose, the requirement for leadership will diminish.

The characteristics of group decision making have been classified in a number of ways. Perhaps the most influential classification is that of Vroom and Yetton (1973) who were concerned with the degree and type of participation encouraged by the leader. In their view, *autocratic* decision making takes place when a leader uses a group to produce all the data required for a judgement, but does not share the nature of the problem and goes on to decide alone. Important, well structured problems which are heavily dependent on a range of expertise are often made this way. *Consultative* decision making involves sharing the problem with the group and discussing options prior to making a decision alone. These decisions may be of a similar nature to those mentioned before but the leader may feel that group involvement is essential to ensure acceptance and to diminish potential conflict. Where the decision is of lesser importance, and acceptability and motivation factors are important, *participative* decision making is used. This allows the leader to share with the group the nature of the problem, the generation of alternative solutions, and the process of arriving at a mutual agreement concerning the solution.

Sundstrom *et al.* (1990) use two dimensions to describe group types—differentiation of members and the degree of organizational integration. For example, project management teams are highly differentiated because of the wide range of skills required, and may or may not have a high degree of organizational integration. Such highly differentiated teams require high levels of information to be processed and integrated, and they are the primary focus of this chapter. These groups must communicate effectively.

Sinclair (1992) has pointed out that groups with a reputation for decision making may not be effective in their greater organizational role. In such groups, real participation is small and the group morale is often low. Decisions are forced on the group by coercive means. In contrast, groups that appear to their participants to be hard working are often characterized by excessive and unreal collaborative exchanges. Clearly group performance is a function of whether you ask the participants, outsiders or use some more 'objective' measure. Certainly, the number of decisions is a poor measure. However performance is measured, certain factors are always important (Kleindorfer *et al.*, 1993; p. 215).

- The nature of the task itself.
- The composition of the group in terms of heterogeneity/ similarity of personal characteristics and backgrounds.

- The agenda the group follows.
- The interrelationships among the group members.
- The degree of power held by each individual.
- The behaviour of the group leader.
- The time pressure and incentive structure under which the group operates.
- The resources available to the group (e.g., group members' knowledge, availability of data, computers, etc.) for its task.

Hirokawa (1992) emphasizes the importance to effective groups of sufficient good quality information. This data is at the centre of all group interactions. He found that effective decision groups use all the available group knowledge to arrive at an agreed problem definition and to discuss their goals and alternatives. At this point the group will plan their work to minimize the collective effort required to arrive at a solution. This is followed by extensive discussions about the positive and negative aspects of the alternatives under consideration. At all stages the effective group will gather information from a variety of sources which will be evaluated for quality and relevance.

And good communication is important to group effectiveness. In tests, where the leader encourages participation in discussion, groups tend to out-perform most individual group members. However, it should be noted that groups rarely do better than the best member of the group working alone (Plous, 1993). Moreover, after reviewing the literature, Hill (1982) has concluded that it is the communication of cues generated independently that is the most important aspect of effective interaction.

He concludes that

- for easy tasks, large groups are more likely to contain someone who knows the right answer
- for difficult tasks, the advantages of groups comes from the pooling of resources and mutual correction of errors
- the best way to generate alternative solutions to a difficult

problem is for each member of the group to work on it independently and to pool their ideas in a meeting.

Some form of debate and even conflict about the data contributed is important. Merely aggregating produces a result inferior to dialogue. Above all, the leader must first concentrate on achieving a shared understanding of the problem. Without explicit agreement about the true nature of the problem, progress will be painfully slow (Orasanu and Salas, 1993; Ellis and Fisher, 1994).

The procedural conduct of the group leader is also important. Research by Korsgaard *et al.* (1995) has confirmed the commonly held view that group members must perceive that their views have been heard and taken into account by the leader. If they feel ignored or their opinions are obviously disregarded then they will consider the process unfair and are unlikely to commit to the decision. Procedural fairness and consideration are particularly important if the opinions of some group members must eventually be disregarded. It is also important that the group leader does not allow conflicts of judgement to transform into personal conflict. Whereas disagreements concerning the nature and weight of cues tends to enhance the quality of decisions, emotive personal disagreements tend to adversely affect decision quality, decision understanding, and group commitment (Amason and Schweiger, 1994).

Team decision making

Much of the research on group behaviour has been done in laboratories using *ad hoc* groups of students. Hardly representative of real groups. This is particularly so when we consider teams —groups that have worked together long enough to develop a degree of mutual understanding and unitary action. Luckily the USA Armed Forces have been studying real team decision making for some years. We will follow the description of this work by Orasanu and Salas (1993).

Shared mental models
This concerns the ability of a team to share cultural, role and situation specific knowledge. Thus members of a project management team may share a great number of social assumptions, an understanding of the corporate culture and ways of operating. Moreover, through repeated interaction and socialization, they soon become aware of the personality and protocols of other team members. Thus, under normal circumstances the team can function in a more or less unitary fashion without misunderstandings. In novel situations the team develops shared situation models (as an extension of the more general mental model) to suit the specific problem. Communication between members indicates the situational response and strategic cues needed for future action. Thus, when a new problem arises less time is spent negotiating, and team members can volunteer information or take action at appropriate times. Other team behaviour becomes predictable. For example, Orasanu and Salas noted that teams that became lost in wartime depended on the leader quickly building a shared mental model by communicating the situation assessment, setting and maintaining goals, and regularly communicating status updates. It has also been observed that superior aircrew performance under stress is associated with 'more commands and suggestions by both the captain and first officer, more statements of intent, more exchanges of information, more acknowledgements and more disagreements' (Orasanu and Salas, 1993; p. 333). They built shared mental models for future use which, with experience, resulted in communication in a sort of shorthand of highly conventionalized language.

Experienced teams develop a sort of *group mind* analogous to memory. Thus, all members remember what knowledge resides with each member of the team through a set of often idiosyncratic labels. This sort of information is used by teams, as different as military command and control units and cockpit crews, in a manner similar to that described in the individual recognition-primed decision model. Thus the team will use their experience to recognize and classify the situation, pick the most likely response and use mental simulation to evaluate its adequacy. In other words, experienced teams tend to make decisions as if they shared a single mind. It is stressed by Orasanu and Salas, that for this response to develop the team must actively communicate at all times during practice and during the real thing. The shared model building must be continuously reinforced by talking of situation assessments, plans and roles. Above all, the whole team must share the same problem definition so that each member can interpret their own roles from their own perspective and, at the same time, be predictable when action is required.

Constraints on group decision making

Groups are not free of the psychological difficulties that beset individuals. Groups, like individuals, attribute success to the group and failure to circumstances, and groups use heuristics (and suffer from their biases) in a very similar way to the solitary judge (Plous, 1993).

Communication is at the heart of good group decision making. However, as we all know, misunderstandings are common between even close colleagues. You may think you understand the goals of the group and its proposed methodology, but this may be an illusion. Your early false assumptions may persist despite contrary evidence and be bolstered by the illusion that other team members share your assessment. The group as a whole may understand the situation incorrectly and group conformity pressures will diminish the doubts of individual members (Orasanu and Salas, 1993). Factors such as these are compounded by the common human failings arising from the misunderstanding of messages. Sometimes, interpretations of instructions need only be slightly wrong for disastrous consequences to occur. The literature on human error is replete with examples.

Orasanu and Salas also point to the errors of group leaders—particularly when they hold power over the group members. They may not hear messages from subordinates if they are not expressed with sufficient urgency and force. As status adds credibility, a leader may easily induce compliance with an incorrect assessment or solution. In a similar manner, it has been observed in flight crews, that the opinion (sometimes correct) of the second in command holds little credibility with the crew compared to that of the captain. This could be one reason why disciplined teams that have learnt sufficient trust to act in an open democratic manner out-perform teams dependent on autocratic leadership.

Groups conform

The advantages of working together and developing a team behaviour are considerable—particularly when multiple expertise is required to generate diverse cues for judgement. But the closer a group grows to unitary action the greater is the danger that conformity will lead the group to do things that individual members would not. The frightening scale of this propensity to conformity is illustrated by a classic experiment performed by Asch (1956). He asked a sample of people to compare the length of three lines (A, B and C) to a reference line. The answer was very obviously B and only 1% of the sample judged incorrectly. However, when other subjects had to wait in a group and the first three were primed to say A, the error rate for the next real subject jumped to 33%. In other words, one third of those that followed were susceptible enough to group pressure to pick a quite obviously wrong line. However, if after six people said A just one said B the error rate dropped to 6%. This firmly demonstrated the value of the nonconformist in a group in search of the truth. On the other hand, the influence of a small number of persistent group members on the majority was demonstrated by Moscovici *et al.* (1969). When judging between blue slides of varying intensity the subjects were told that it was a test of colour perception. However, when a small number of primed people confidently judged some of slides to be green, 32% of the real subjects judged at least one of these slides to be green. Thus, the nonconformist can influence the majority negatively as well as positively.

Groupthink

Teams working under stress with a strong leader may have a problem that Janis (1972) called 'groupthink'. Ministers and their minders, generals and their staff, Presidents and their cabinets are the sort of examples that Janis used. Thus, the obviously mistaken Bay of Pigs invasion of Cuba in 1962 by the USA could be seen to be a result of groupthink. Early in his term, President Kennedy had surrounded himself with like-thinking advisers. Under pressure to demonstrate strong leadership, Kennedy requested advice from this group and was only given a few options to consider. This inner cabinet shared a sense of invulnerability and virtue, tended to demonize the opposition, demanded loyalty from all cabinet members and censored any doubts. Thus, counter arguments and inconvenient intelligence was ignored. The outcome made a fool of Kennedy and, as a result, he made sure that, in the future, he was advised by a wider group. This more diverse group was forced to debate the issues without presidential intervention, and only after receiving opposing views would Kennedy make his decision.

It is unlikely that groupthink behaviour is confined to the political agenda. At the top of some business and government organizations there are probably chief executives who surround themselves with sycophants anxious not to displease the boss. They, too, would be secretive and censorious of inconvenient information—acting as mind guards and gate keepers to maintain their privileged positions. In fact, any boss would be wise to be uneasy if too many top managers and special assistants are saying 'yes' too often.

Groups polarize opinions

If a group is debating the virtues of two solutions to a problem, and two more or less equal factions develop, the result is likely to be a compromise. However, if a majority have an inclination one way, the end result is often a more extreme version of that solution

(Stoner, 1968; Moscovici and Zavalloni, 1969). In these circumstances, people come out of a meeting having supported a more radical view than they would have taken prior to the meeting. Similarly, prejudice is increased after discussion between like minded people (Plous, 1993). Kleindorfer *et al.* (1993) have canvassed a number of possible reasons for the polarizing effect

- the majority say more
- people in groups tend to stand back and let others take responsibility
- some people want to be seen to be more forceful than others
- the majority opinion is legitimated by denigrating the minority opinion
- it may sometimes be the result of the way the problem is framed.

Whatever the reason, we should be careful about the ability of groups to take risks that would not result if the opinion of individual members were canvassed one by one.

Managing group decisions

In this section we will again use the model of the decision process shown in chapter 7 but with a major modification. Communication must be added if we are to reflect the core process in group decision making. Without communication no collective action can be taken. The stages will now look like this

- problem or opportunity recognition
- communication
- thought (search and quasi-rational inference)
- judgement (end result of inference)
- decision (intended action)
- action.

The only other necessary difference between individual and group decisions is that two or more people contribute to any of the other subprocesses. It is most likely that they will be required to contribute new data in the search process and perhaps contribute insights to the inference derived from that and other data. Moreover, the next chapter will discuss network decisions that involve large numbers of individuals, groups and organizations in the decision process. Therefore, to differentiate between groups and networks we will assume that the groups are of meeting or project management team size.

We will now proceed to look at the management of group decisions in or out of your professional domain.

Circumstance I.
Decisions in your personal professional domain

First response
Always make a recognition-primed decision first.

- *Problem recognition.* You will recognize the problem as one within your personal domain of expertise.
- *Communication.* Let the team know that you may need their input.
- *Thought.* Think of the possible causes and likely problem development.
- *Communication.* Bounce your ideas off the team and modify as necessary. Encourage devil's advocates but discourage lengthy discussion.
- *Judgement.* Use your experience to choose a typical, practical, solution option.
- *Communication.* With the team, run through the likely consequences of adopting this solution. If everything is not OK repeat the *judgement* stage.
- *Decision.* If everything is reasonably satisfactory with your mental simulation, adopt this solution.

- *Action*. Keep observing the situation for new or changed cues that will require a change of plan. If the new information makes your chosen solution untenable, abandon it and treat the situation as a new decision opportunity.

Second response

As we have noted before, a recognition-primed solution may satisfy the goals but will not be optimal. If you are required to demonstrate an attempt at optimality, put this recognition-primed solution aside as a back-up and proceed with the use of a combination of the reasoned choice and image theory models.

- *Problem recognition*. Ask your team members to independently list all likely interested actors and their likely goals and problem definition.
- *Communication*. Meet and discuss their work with the aim to provide a comprehensive definition. Do not reveal too much of your own thinking.
- *Thought*. Assign members of your team to gather data on the attributes of the problem. Ensure that the most important technical, organizational, social and personal factors that are likely to influence choice are covered. This information may be used in two ways. Firstly, it should be clustered into (for example) five categories for ease of judgement using the image theory procedure. Secondly, you may assign an expert member of your team to use the data in an analytical decision method such as decision analysis or SMART. Think carefully about the professional and organizational values involved, the goals of the decision and how it may be implemented.
- *Judgement*. Compare options, as described by the data clusters, to the values, goals and plans. This is a holistic, judgemental process heavily dependent on the mobilization of your personal expertise in the domain. Use any decision analysis results only as a check and review them

if they are counter intuitive. Remember your expertise is probably a better guide than the calculation.

- *Communication.* Go over the results with your team to look for inconsistencies and threats to implementation.
- *Action.* Implement and monitor.

Circumstance II. Decisions outside your professional domain

The great advantage of group decision making is the ability to handle problems outside of your own personal professional field. All you need is expertise in management and the right team.

- If your team contains an experienced expert in the field being considered, tutor him or her in the art of recognition-primed decision making and arrange for him or her to go through the steps of Circumstance I (First response) above. Hold this solution as a check or backup.
- Persuade the same expert to help you work through the steps of the more comprehensive analysis detailed in Circumstance I (Second response) above. Your management skills should be concentrated on ensuring that team cooperation is not degraded by your lack of domain knowledge. Keep the goals in front of the team and ensure that organizational values are not violated. Above all communicate and reward, communicate and reward. In other words act like a leader who has the trust of the team and, in turn, trusts them.
- If your team does not contain an expert in the relevant domain, get one. If you cannot, be prepared to manage the consequences of an inferior decision.

These steps as described on the last few pages, seem cumbersome when written down but, in practice, the process is quite swift. Recognition-primed decision making is an action decision process and can take only minutes to perform. The more comprehensive method takes longer. Much depends on how long it takes to gather the decision data. After that, the experienced decision maker will find that the holistic judgement process required by image theory can be done quickly. However, when operating outside of your domain of knowledge you should take account of the communication time consumed in team meetings.

Group decision management—an illustrative case study.
'The Takeover'

'I have a dread,' he said, 'that we have moved too fast.'

'Old dead-beat' thought the MD. The chairman went on to elaborate on his uneasiness.

'We may not meld despite our mutual enthusiasm. How are we to reconcile Phil's disregard for anything other than technical excellence with your obsession with the bottom line?'

'Damn it Jack, stop messing about. We have come too far to back out now.' The chairman expressed his irritation with a sharp turn away from Coulson.

'Bill, your drawback is that you don't know the danger of a bullheaded disregard of different corporate cultures. You seem to think that rationality is defined by money. That money is what we all work for. Well I hope you are not in for a shock.'

Coulson's tense stillness drew the chairman round again.

'Christ,' he thought, 'Phil has resigned.'

It, therefore, came as a relief when the MD described the shock shown by Phil when the financial status of his firm was tabled at the last management committee meeting. It was a very chastened chief of design who agreed to work with the other division heads to try and pull the firm back into profits. But things had gone wrong almost immediately. Phil had shown an almost complete disregard for team working principles. In particular, he had shown his contempt for anything to do with marketing. As he said,

'Those that want the best will come to us. We don't need to crawl to the market place.'

Clearly, the sobering effect of the accountant's figures had faded. For too many years he had been the key to the company's reputation for excellence, and damn if he was going to change his way of operating to please the owners.

'Surely,' Phil thought, 'I am the reason they had bought the company! And I have the backing of the chairman!'

Despite his relief, Jack was deeply concerned about what Bill was telling him. Bill had solid experience working with teams and they both knew how damaging a maverick could be. But Phil was too important to be ignored.

'Our tactics are all important now Bill. I know I am supposed to be above the fray, but you are getting my advice anyway.' The chairman leaned forward, and Bill straightened a fraction in his chair sensing that this advice would be ignored at his peril.

'First of all split Phil off from the management committee and deal with him separately. Teamwork isn't the be all and end all. Dialogue and data are the important things. Explain that you want his advice free of the constraints of the committee room. He's intelligent enough to know he is being destructive. Make sure that everyone understands and agrees with the global problem and ask each division head to submit his views, backed by evidence, of what the specific problem is. Phil is a good analyst and he will be a great help sorting out the wheat from the chaff.'

'Before you go on, Jack, could you please make sure that Phil does not continue to have open access to your office on this issue. It's irritating the whole team.'

'Yes, you're right. But let's get on. Having got everyone to agree, including Phil, about the nature of the specific problem you should bring me into at least some of the meetings. Make sure Phil attends those meetings, Bill. I can keep him in line, and he and the rest of the team will see that you have my backing on this issue.'

'But Jack, you seem to want to spend a lot of time avoiding answers.'

'Well why not? I don't want a solution set in concrete before all the options are canvassed. Keep the folks focused on solving the *right* question, Bill.'

'OK, then what?' Bill sensed Jack's irritation. No one liked to take advice—especially from the boss.

'Well, having agreed to the specific problem, ask each member of the committee—including Phil—to present options and kick them around. When the options are adequately understood, ask each division head to put forward written arguments for and against the options, and then let the committee kick them around a bit. The next move is up to us Jack. You and I have to show leadership by sorting out a solution from the options that will be best, overall, for the new joint firm. We are sure to get plenty of partisan opinion but it's up to us to think of the benefits to our new common culture. Get to it Jack.'

Discussion

The chairman's tactics are a good response to the situation. By concentrating on utilizing the advantages of group work (this is not a team) to gather data, define the problem and present options, he is maximizing both the range of cues and the likelihood of group commitment to the solution. The chairman also recognizes that some people work best outside of groups and, if properly managed, can be an antidote to groupthink. The chairman will, no doubt, lead in the formulation of the final judgement because he has a broader and deeper sense of the cultural differences to be bridged than the managing director. He knows well that the history of mergers is not good. Too often the financial advantages completely mask the human problems.

Summary

- Group decision making in organizations, although slow, has the advantage of contributing a wide range of cues for judgement, improves decision commitment and conforms to current norms of corporate behaviour.
- Groups that have worked together sufficiently become teams which tend to share mental models that guide decision making behaviour.

- Groups may develop certain social pathologies which often result in a flawed decision process.
- Finally, recommendations are made on strategies to manage group decision making.

Part IV

Strategic decision making

This part of the book moves into the complicated realm of decision making involving multiple actors and organizations. This requires the resolution of complex communication interactions, the alignment of interests and the use of powerful manoeuvres to achieve acceptable outcomes. Chapter 10 explores the empirical work that has been done on strategic decision making. The subject has proved resistant to on-site investigation because of the large number of people, groups and organizations involved in the process, and research has therefore depended on written or verbal records. Nevertheless, a picture emerges of the simultaneous use of a number of strategies which may arise from technical analysis or from organizational or personal forces. Two models are presented that reflect this complexity. Chapter 11 is largely devoted to a detailed explanation of how one of the models described in chapter 10 can be used to manage the strategic decision process. Chapter 12 concentrates on certain practical procedures that enable the manager to realistically define the nature of the strategic problem or opportunity, detect possible solutions, and construct an argument for or against a solution. These are important steps in the decision process and it is clear from the literature that a solution that is both effective and socially acceptable can only be achieved if great care is taken in the modelling of all stages. Finally, some concluding remarks re-emphasize the superiority of a *holistic* decision philosophy that uses analysis as only one potential source of data.

10

Strategic decisions

Strategic decisions may be divided into two subgroups. *Organizational decisions* are those that involve multiple levels and diverse divisions of an organization, and are often long term and important. The final decision is often taken by the most senior executives in an organization, but only as the result of many other intermediate, lower level decisions taken over many months or, sometimes, years. They are rarely easily retraced because of the influence of a network of interactions between actors which often produce unanticipated outcomes, disruption and iteration. *Social decisions* are similar in nature to organizational decisions, but always involve people and institutions outside the organization as important actors. The consequences to society as a whole may be more important than the consequences for the individual organizations involved. These decisions require interactions between a dynamic network of actors most of whom are only partially controllable by the decision maker.

Understanding, in these circumstances, comes not from a knowledge, however flawed, of human psychology and behaviour, but rather from a sociological observation of the interactions and variables. Only parts of the patterns will be discerned and only at certain times. Nevertheless, good detective work has produced a respectable body of knowledge.

Multiple perspectives

Graham Allison (1969) was an early but influential organizational sleuth who examined the voluminous records of the decision making process involved in the Cuban missile crisis. You may recall that the USA was thrown into a state of shock by the discovery in 1962 that Soviet missiles were in place on the island of Cuba. For 13 days the USA was on a war footing, and the confrontation with the USSR was so serious that President Kennedy estimated the chances of nuclear exchange to have been between 33% and 50%. It could be said, therefore, that the strategic decisions taken during this period were the most important in human history. From our perspective, the importance of Allison's work lies in the generalizable nature of the results. The picture that emerged explains not only important sociopolitical decisions but also the interactions of many large organizational networks of actors.

It is often assumed that strategic decisions are the result of the efforts of a group of close associates at the top of the organization deliberating on the results of a rational analysis of the available data. Allison found that this *rational policy paradigm* was indeed used—particularly by the intelligence community—to make sense of the situation and provide action options. This paradigm embodied the base assumptions of political analysts of the day. The key organization, in this case the US Government, is assumed to be a rational, unitary decision maker which responds to a clear problem by defining its goals, identifying options and their potential consequences, and choosing that option most likely to achieve the goals. However, Allison found two other rationalities at play in the decision making. The first, he called the *organizational process paradigm*, does not assume a unitary actor but looks upon decisions as arising from the routines of loosely allied organizations. The problem and the relevant data is cut up and divided between the organizations in accordance with their expertise. Thus, goals become parochial and addressed in order, using standard operating procedures and well rehearsed programmes. Central coordination of the organizational outputs becomes difficult and responses inflexible.

The other influence Allison called the *bureaucratic politics paradigm*. This is the dynamic bargaining for influence that takes place between powerful actors. Goals become interests, and outputs the result of bureaucratic political games played in accordance with the structural constraints of position and power. Thus, parochialism also dominates this paradigm and the end results, when the power plays are resolved, are often unpredictable. The indian that wins is the indian that grabs the chief's attention.

Allison's (1969) study has gained its reputation because it fits so well the experience of many players in the organizational game—we recognize these elements in the flux of organizing. Nevertheless, the cake can be sliced in other ways. Linstone's (1984) review of the literature demonstrates that in other circumstances the two organizational processes identified by Allison may be usefully collapsed into one. Other circumstances may require an expansion of categories to differentiate between (for example) small group processes, the role of a dominant leader and the key effects of the highly flawed human cognitive process. In all cases, however, some form of the rational policy paradigm is noted.

Linstone (1984) is concerned with decision making involving *sociotechnical systems* where the technology is important but we must, nevertheless, take account of the human and social factors surrounding and interacting with it. He considers that three perspectives are adequate for analysis. The *technical perspective* is similar to Allison's rational policy paradigm but explicitly includes quantitative analysis. Models, data analysis, probability, decision analysis, economics and engineering design factors would be typical of the tools used in this perspective. Linstone's *organizational perspective* includes both Allison's organizational process paradigm and his bureaucratic politics paradigm but expands the organizational actors to include groups external to the central organization—including public interest groups. This perspective is of help in sociotechnical decision making as it enables

- identification of the pressures in support of, and opposition to, the technology
- insight into the societal ability to absorb a technology-organizational incrementalism as an important bound

- increasing ability to facilitate or retard implementation of technology by understanding how to gain organizational support
- drawing forth impacts not apparent with other perspectives, for example, based on realities created within an organization
- development of practical policy (for example, new coalitions) (Linstone, 1984; p. 52).

Finally, Linstone suggests that our understanding will be enhanced by the *personal perspective* which identifies the role of charismatic leadership. He also points to the illumination of the problem and its dynamics that can only come from interviewing key individuals involved. Naturally, a full understanding of the political manoeuvres within organizations can only come about by considering the details of individual behaviour and motivation. Linstone goes on to illustrate how these three perspectives could be used to map the dynamics of a wide range of sociotechnical problems and how the relevant decisions came about.

The nature of the problem is also an important influence on the decision process.

Problem type and complexity

In what has become known as the Bradford studies, Hickson *et al.* (1986) examined 150 case studies of strategic decision making from 30 organizations, large and small, and across the private and public sectors. Their aim was to explain the process of decision making in terms of the problem types and the organizational interests affected. The 150 cases involved infrastructure (23), reorganizations (22), corporate planning (19), marketing (18), expansion (16), new products (12), personnel (12), mergers (11), finance and resources (9), and organizational location (8).

Decisions were analysed in terms of their complexity, progress over time, internal political importance and the organizational interests involved. A combination of these factors indicated that decision making had three dominant modes

- *Familiar.* These were topics often experienced in the organization which contained little complexity and were relatively unpolitical. These decisions tended to progress in a channelled manner such that only a limited number of managers were involved. Routine, expert procedures were used, with the final decision in the hands of middle management.
- *Unusual but tractable.* These were matters of intermediate complexity but, because they had little political content, tended to move through the organization in a fluid way. They were formally channelled through committees, had less expert involvement, and the final decision was taken at the highest level.
- *Weighty, controversial, 'vortex'.* These matters tended to involve many internal and external interests and, therefore, were highly political in nature. They involved the utilization of a wide range of data and expertise, extensive formal and informal interaction, and experienced frequent lengthy delays. The final decision tended to be made at the highest level.

It was concluded by Hickson *et al.* (1986) that, whatever the type of organization, it was the *subject matter* of the decision that most affected how it progressed through the organization. Some subject matter did not involve all the modes. For example, decisions about products were either familiar or vortex in nature but those about infrastructure could be either familiar, tractable or vortex. Thus, the Bradford research indicated that, whatever the organization, big or small, public or private, decisions about personnel, mergers and finance are likely to cause similar problems, activate similar interests and progress in a similar way. The very nature of the subject matter moulds the process of decision making. Thus, we may conclude that engineering managers, coping with many vortex decisions, may experience the decision process in rather a different way to an accountant dealing with financial matters of a familiar or tractable sort.

Whatever may be the role-rationalities associated with marketing or production, personnel or finance, Hickson *et al.* (1986) consider them all to be subject to two other levels of rationality. At the highest level is the *rationality of control* that defines the norms, culture and regulations of the organization. Within the constraints of this rationality may be found two other. The *rationality of problem solving* structures the complexity of process in a more or less rational manner and the *rationality of interest accommodation* copes with the political nature of the situation.

McCall and Kaplan (1985) have remarked how strategic decisions are comprised of smaller decisions, each influencing the other. Thus

1. While there are discrete decision points, it makes more sense to look at streams of decisions. Decisions are really accumulations of subdecisions, usually made over long periods of time.
2. Decision making is sometimes an individual event, but more often many people are involved in various ways and at various times. This is another reason to look at decision streams—problems as they flow through organizations—rather than individual decision makers.
3. Decisions do not unfold in logical orderly stages. They double back on themselves, solutions are found before problems are understood, action on earlier problems affects current decisions, and so on. Because of this interwoven quality, problem solving and decision making are not separate activities (McCall and Kaplan, 1985; p. 104).

With this process in mind, Kriger and Barnes (1992) examined 147 decision events in two large American manufacturing companies and concluded that six nested decision levels could be detected.

- *Level I: decision choices.* These comprise moments of resolution such as a final bid, an offer acceptance or the decision to speak to a tardy employee—the sort of potential actions that can be represented in a decision tree. They are the obvious and identifiable choices in an ongoing decision process that round off a subprocess and initiate another subprocess.

- *Level II: decision actions.* These are the implementation of several level I choices at meetings or through the drafting of letters or memos. Level II decisions make known level I decisions.
- *Level III: decision events.* These are the wider communication of level I decisions, taken over days or weeks, that involves a number of level II decision actions and a wide network of actors both inside and outside the organization.
- *Level IV: mini-decision processes.* These are the identifiable sets of level III events that culminate in some strategic action. These mini-processes often take several months or a year to conclude. A company acquisition decision is an example.
- *Level V: decision processes.* These comprise identifiable long term strategic decision processes comprising a number of level IV mini-processes accumulated over many years. The long term strategy to enter a new market would be a suitable example. This is very much a historical modelling of other levels of decision and, as a result, the corporate memory of the thousands of past level I decisions is often partial.
- *Level VI: decision theatres.* These can be discerned from the long term history of a company or state. The strategic decisions of the USA concerning Vietnam is an example.

The usefulness of this research lies in its demonstration that what is commonly known as decision making is often only one stepping stone in a long term process involving a steadily widening network of actors, long time periods and a high degree of communicative complexity. The cyclic nature of the processes is also implied in this research. A chief executive may initiate a level V decision process after long deliberation which can only be realized by thousands of level I decisions which, in turn, aggregate into the other levels of decision making. Alternatively, incrementalism in decision making implies that what may appear to be a strategic decision may be only the inadvertent result of an accumulation of strategically unguided decisions at levels I and II. True strategic decisions are the result of a response to a high level problematic situation, and initiate the other levels of decision making. Nevertheless, strategic decisions are themselves heavily influenced by prior decisions, or the nested accumulation of previous decisions, which may or may not have been strategically guided.

Organization structure and context

Not only the nature of the problem affects the strategic decision process but also the nature of the organization within which the decision occurs. Frederickson (1986) considered that three dimensions of organizational structure were important to organizational decision making. *Centralization* measures the degree that organizational decision making is concentrated at the top. *Formalization* indicates the degree to which decisions and other organizational factors are moulded by standard procedures and rules. *Complexity* refers to the horizontal and vertical differentiation and/or spacial dispersion of an organization.

A centralized organization enables active goal-focussed decisions to be made that may be radical departures from current procedures. Thus, organizational decision making can utilize what Allison (1969) would have called the rational policy paradigm, such that carefully selected means can be directed towards the achievement of agreed and stable goals. The drawback of centralization is the limited search-inference procedure which may result from limited participation. The data and inferences used in the decision making will be enhanced in less centralized organizations, but increased goal conflict will encourage more parochial and incremental decisions.

Unlike the proactive, opportunistic decision making of the centralized organization, an organization that has a high degree of internal regulation tends to be reactive in its methods. The goals of the decision makers are often limited and measurable but, nevertheless, the dominant rationality ensures that it is the means that dominate decision processes rather than ends. Thus, Allison's (1969) organizational process paradigm is a good description of the dominant mode of decision making in these organizations.

Complex organizations are rich environments for the potential formation of strategic initiatives. However, the parochial concerns of highly differentiated or dispersed groups mitigate against the pursuance of opportunities that are not in the direct interests of those groups. Directing decisions is also difficult because of the high degree of unit goal conflict which mask the agreed global goals. Therefore, decisions tend to be parochial and incremental and result from the resolution of internal political conflict. It is clear that Allison's (1969) bureaucratic politics paradigm fits this type of organization well.

These dimensions may be found to some degree in any real organization and the resulting interaction between decision making styles will be complex. Nevertheless, Frederickson's work is a useful indicator of what type of decision process is likely to dominate in organizations as diverse as government departments, large public corporations or small entrepreneurial firms.

Other aspects of the environment affect decision procedures. Dean and Sharfman (1993) examined 57 strategic decisions in 24 companies to reveal to what extent the rational search-inference procedure was approximated. In other words, what influenced a company to use the rational policy paradigm. They found that, when the level of competitive threat was high, the use of rational procedures was low and the importance of the problem had no effect on the rationality of method. Procedural rationality was highest when the organizational locus of control was internal—when external stakeholders were unimportant and when external threats were limited. The decision process was also more likely to use the rational paradigm if the level of uncertainty was low.

These findings tend to confirm the view that cool rationality is a rarity in most organization contexts and that situational constraints make optimization difficult to achieve. Only in the rare circumstance that external and internal pressures are low and stable will the management team be able to respond to problematic situations by coolly gathering information, analysing it using rational procedures, and making choices which maximize utility. Usually, external and internal pressures are such that choices have to be made on the run with limited information about the current situation or future consequences.

Effective strategic decision processes

Whatever it is that determines the decision process used, there is some evidence that effective processes have a particular set of characteristics in common. Harrison and Pelletier (1993) looked at the records of 24 very large strategic decisions made in some of the biggest USA businesses. They found that the most successful decisions involved the following characteristics

1. The objectives of the decision process were well defined and attainable.
2. The decision process was highly interactive and open to many views. In particular, alternatives and outcomes were not predetermined at the start.
3. No attempt at optimization was made. Consideration was given to only a number of feasible alternatives and the decision process was closed when the objectives were achieved.
4. The decision maker accepted the uncertainty and ambiguity of the situation and used a judgemental rather than computational strategy.

Whenever any one of these characteristics were not evident, the decision was less than effective. The worst decisions resulted from a process that involved a closed set of interactions within the organization, the generation of unattainable goals, and an attempt to maximize some outcome by computational means. Although it would be rash to generalize too much from this limited piece of research, the findings tend to conform to what is known about more limited organizational decision making. Certainly, a process such as that described in image theory echoes these four criteria for success, as does the power model that follows.

Two models

The Mintzberg et al. *(1976) model*

This model of decision making was based on the study of 25 Canadian strategic decisions. A simplified version of the model in diagrammatic form may be seen in Figure 15.

- *Recognition.* This occurs as the result of both seeking opportunities for action and as a response to problems or crises. This study found that the vast majority had some element of problem recognition as a stimulus for decision making action. It also appears important that some hope of a solution is evident before a manager will initiate the decision process.
- *Diagnosis.* The use of available information to clarify and define the issues. This step may not be discernible in some decision processes—particularly those few that did not involve a problematic situation.
- *Search.* This is evoked to find ready-made solutions in a process moving from a search of available memory sources to the use of active scanning procedures. Screening out of unsuitable alternatives takes place at this point.
- *Design.* Used to develop custom-made solutions. This is a complex iteration procedure which, in effect, works through a decision tree to narrow the alternatives to one or two potential solutions. In many cases only one potential solution is identified prior to moving to the next

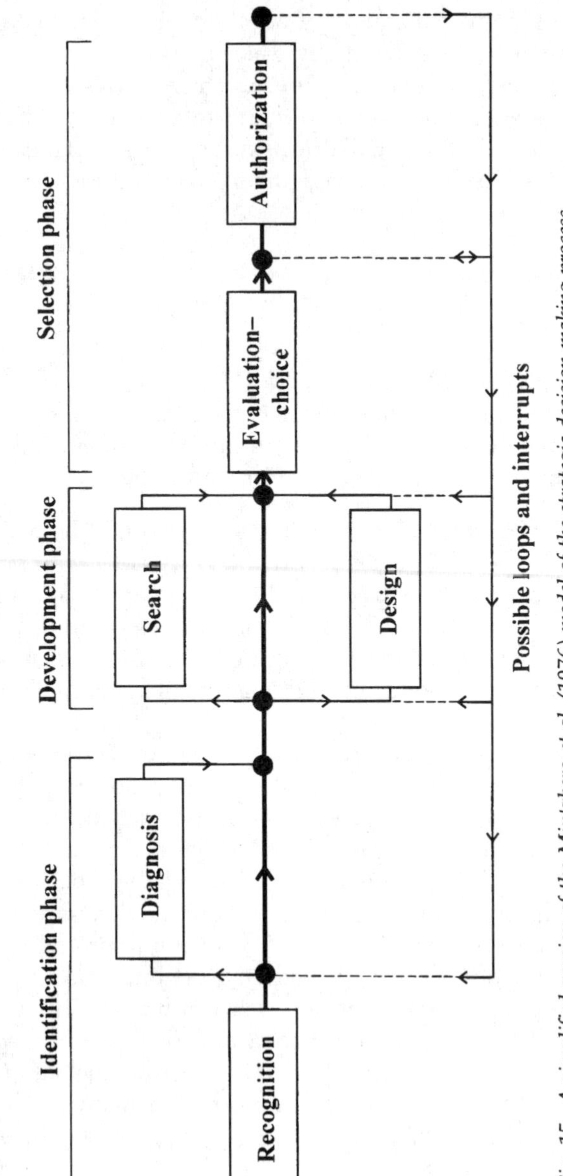

Fig. 15. A simplified version of the Mintzberg et al. (1976) model of the strategic decision making process

stage. Design and search are often found to be parallel or nested, and these two protocols absorb the greatest amount of time and energy in the whole process.

- *Evaluation–choice*. This was found to be a less important phase than the normative literature would suggest—at least compared to diagnosis or design. In only a small number of (largely technical) cases could procedurally rational evaluation be detected. In the vast majority, evaluation and choice were intermixed, with considerable iteration and re-evaluation taking place. Evaluation–choice was achieved using individual judgement in some cases, but in the majority of cases after a process of bargaining between individuals or groups that had arrived at different judgements about the alternatives. Even when technical analysis was performed, the final choice was made using judgement and bargaining. In no case was any form of decision analysis used.
- *Authorization*. The process of seeking formal approval for the implementation of the chosen solution. This stage exposes the problem and solution to new actors with a wider agenda which often results in the abandonment, reprocessing or modification of the problem/solution.

It was noted that activity frequently looped back to a preceding stage and numerous interruptions to progress were observed. Mintzberg *et al.* (1976) also emphasized the importance of routines that control the direction of the decision process and manipulate and disseminate information, and the pervasive presence of political bargaining throughout the decision making period. The importance of these matters is emphasized in the next model.

The power model (Parkin, 1994a, 1996)

This model was developed using the empirically based actor-network paradigm—sometimes called the sociology of translation—described in chapter 2 (Latour, 1987; Callon, 1986; Law, 1994). It emphasizes the importance of problem definition, the network interactions through judgemental argument and finally, the necessity to minimize the need for coercive power during implementation. You may recall that actor-network theory is derived from recent empirical work done within the sociology of technology. It acknowledges the role of artefacts and techniques in the process of organizing, and generously treats them as important actors in the network of nested interactions that constitute the parts and wholes of organizations. Therefore, this framework is uniquely able to model strategic sociotechnical decisions where engineering and science play an important role.

The power model is diagrammatically represented in Figure 16. The control axis is similar in logic to the model of the individual human decision process shown in chapter 7, with the search process replaced by network interactions. A brief description of the role of the flanking boxes will hopefully illuminate the process.

- *Constraints.* Although actor-networks are made up of a heterogeneous mix of actors, some of which are non-human, it is useful to discriminate between those that can and those that cannot make judgements. For example, technology may shape outcomes in important ways by constraining or enabling options, and the environment may make some solutions impractical, but they are, nevertheless, incapable of wilfully guiding the decision process.
- *Problem definition and enrolment.* This describes the process of moulding the nature of the problem and the capturing and control of other members of the network. Thus, the actor that first reacts decisively to the problematic situation may define the specifics of the problem (or opportunity) in terms favourable to its interests. This key actor then proceeds to enrol other actors into this definition by *translating* their version of problems and their interests in a way consistent with that definition. Maintaining the hegemony of the definition and sustaining the enrolment of the other actors is a skilled ongoing

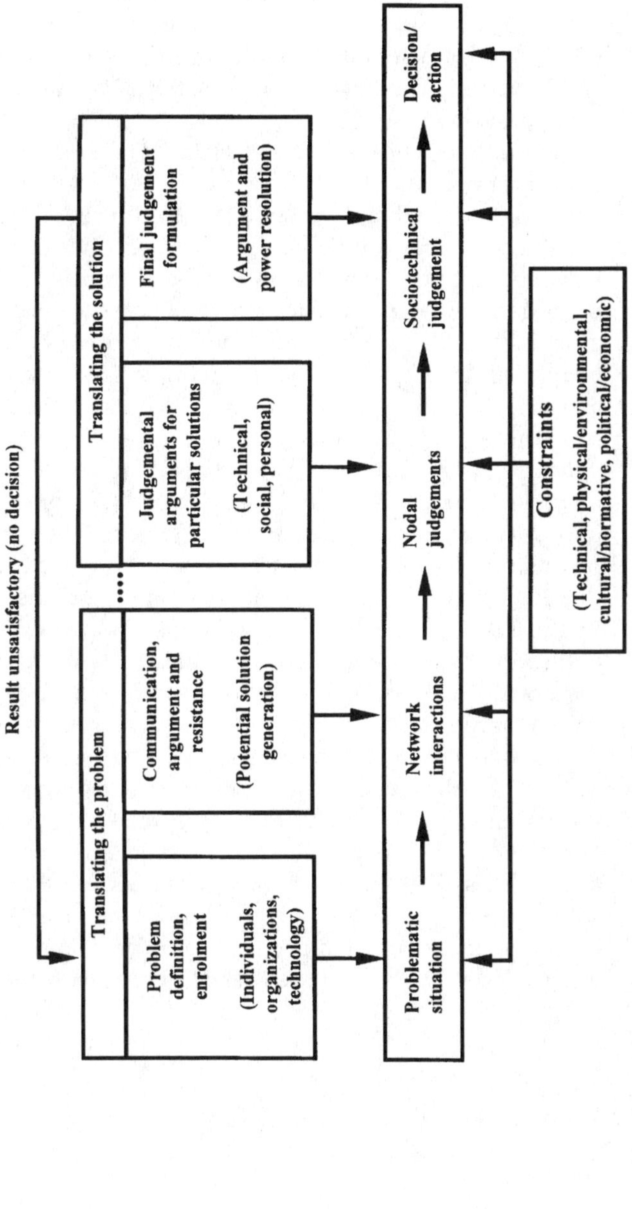

Fig. 16. The power model (Parkin, 1994a)

act of manipulation, and is always prone to failure as other powerful interests compete for control. Part of the enrolment manoeuvre is to bestow legitimation on the other enrolled actors and to take on the role of network spokesperson. Thus one actor is sustained as the 'centre of translation'.

- *Communication, argument and resistance.* This is the formal and informal interaction that takes place within the network to embed and perpetuate the problem definition and the potential solutions. Thus, through this communicative argument process the power of translation and the hegemony of the enroller are sustained against points of resistance. Reports and other written instruments will be used at this stage to make the definition more tangible and thus enhance its durability. Potential alternative solutions are often generated at this stage.
- *Judgemental arguments.* These are the end result of the ongoing network dialogue. Each actor will come to a judgement concerning the nature and merit of the various potential solutions to the problem, (as now defined) driven by their fundamental interests and role language. Engineers will use technical cues for judgement and politicians will use social cues, and the appropriate arguments will be expressed to others in the network and, in particular, to the actor at the centre of translation.
- *Final judgement formation.* This is the evaluation–choice process to decide which solution will be used. As implementation of complex sociotechnical decisions depends on the cooperation of a range of institutions and publics, it is necessary and desirable to choose a solution that will not require undue coercion to enact. Thus, the judgemental argument that is finally chosen or formulated is likely to be one that demonstrates a good fit between the values embodied in the solution and the salient social values. The better the fit, the smaller will be the degree of coercive power required to implement the

decision. If too much coercive power is required the problem may be redefined and the process restarted, or the whole process may be abandoned.

Naturally, the embodiment of a complex decision process in a diagram implies a simplicity that is not evident in real situations. The problematic situation, for example, may have been created by the accumulation of other decisions and any solutions will be constrained by their nature. The dynamics of the network interactions and the fragility of the enrolment may produce a number of shifts of the centre of translation from one actor to another . This is particularly true of the end game, as the initiator may lose control to a higher level decision maker. The whole process, from problem definition to final judgement, represents a resolution of conflicting interests to minimize the use of power to enforce the resulting action. To be well done this needs political savvy of a high degree.

The power model is used as a basis for the chapter on the management of strategic decisions during which the process will be further explained and illustrated.

Summary

- The network of interactions involved in organizational and inter-organizational decision making requires the use of multiple strategies which arise from such factors as the complexity of interests involved, the task characteristics, or the structure of the organization.
- The Mintzberg and power models are presented to indicate the interrelationships involved in strategic decision making.

11

Managing strategic decisions

I have devoted a whole chapter to this subject because of the complexity of the problem and the highly political nature of the strategic decision management process. The recommendations are based on the power model (Parkin, 1994a, 1996) described at the end of chapter 10, and diagrammatically represented in Fig. 16. This model emphasizes that the decision process involves a large number of individual and organizational actors (some of which are non-human), both within and without the organization. These actors are connected by a communicative network which is only partly controlled by the principal actor. To achieve the effective decision outcome that you want, you must be prepared to manipulate and mould this network using persuasive and powerful manoeuvres. You must be prepared to be political. In particular, prior to the decision/action point, two global and subtle acts of translation must be achieved—*translating the problem* and *translating the solution*. These acts of translation are powerful manoeuvres designed to align problems and solutions with interests and value systems. The rest of this chapter will be devoted, to a description of these translation techniques. The first phase, the translation of the problem, is derived from actor-network theory (Callon, 1986, 1987, 1991; Latour, 1987; Law, 1986, 1991, 1992a, b, 1994; Law and Callon, 1992).

Phase I: Translating the problem

Machiavellian manoeuvres

An actor-network may be conceived of as the successfully stabilized interactions and relations between interested individual, organizational and physical entities associated with the achievement of, for example, a new piece of technology, a project, or a strategic decision. Let us now discuss the tricky manoeuvres used to produce and stabilize an actor-network required to arrive at a strategic decision. The key is the translating of interests or problems such that others are enrolled in the decision process and their subsequent behaviour continues to act in the interests of the proponent.

Cooperating in your decision process is in the interests of another actor if it helps them achieve their own goals. Thus, the enrolment of another actor may be simple if your process directly satisfies their requirements, or your interests can be furthered by joining a scheme of theirs. You may even be able to convince others that although their interests cannot be completely satisfied directly by any means, your scheme could represent a possible compromise solution. Inventing replacement problems for other actors may be possible. Thus, you enrol them—leading them step-by-step into alignment with your interests. The desired end result of this machiavellian manoeuvring is that you and your decision process will become an obligatory passage point through which all the network of actors' interests would be satisfied. It would become a centre of control of resources, information and decision making.

Another form of translation of particular salience to engineers is what is called in actor-network theory, 'problematization'. This is a manoeuvre which requires you to convince the other actor that their problem is so similar to yours that they should accept your proposed decision making strategy. As you own the strategy you then become indispensable—an obligatory passage point. Your strategy may even involve a solution. For example, proponents of a light rail system may convince businesses along the route that their staff and customer access problems are similar to those of your client at an entertainment centre, and that all these problems can be solved by the installation of a light rail system. For those that obviously do not share this problem it remains to convince them that the extra clientele will clearly offset the inconvenience of the construction—that their interests correspond to yours.

The role as an obligatory passage point may be reinforced and used to maintain close control over the actor-network. In our example, until the light rail system is in place the wise proponent will attempt to manage all communication, resource flows and decision making on behalf of all interested parties. Indeed, in the process they will attempt to define the roles of all parties and constrain their ability to resist. They will also hope to have gained legitimacy as a spokesperson for all the network actors. It is an important move for the proponent to define exactly the nature of the other actors in the network and to firmly establish their role as representatives of their constituencies. This enables the dominant actor to put in place meetings, briefings, reports and other procedures that block any potential contact with others outside the network and focuses attention on the obligatory passage point. This, of course, is why some public interest groups will refuse to cooperate with government liaison arrangements or public information evenings for fear of being sucked into procedural arrangements that may constrain their ability to act.

I should emphasize that these moves are not the product of a fevered mind but the observed characteristics of actor-networks that have been successfully formed and stabilized in the real world. Moreover, those that have failed to gel have clearly been deficient in some aspect of the translation process.

The nodes of an actor-network are, of course, actor-networks themselves. Sometimes, as the network develops, one of these node-networks may be of such complexity and importance that the internal node interactions are as much an influence on outcomes as those of the greater actor-network. In engineering design and production this is the norm. A global network first forms, in the tentative manner of all actor-networks, and defines implicitly or explicitly its problem in such a way that certain types of solution are implied. Thus, a transportation agency may enrol into an actor-network other agencies, private port operators and a particular run-down sector of the port to provide a new crossing of the river. This global network thus provides both the resources and the protective environment for a local network of designers to sketch, draw, model, make mistakes and generally experiment away from the gaze of the greater public. The global network provides the resources for this design process and gratefully (or otherwise) receives the results as its reward. Failure of the project may occur at any time and in either actor-network. The global network may lose control over resources because of government changes or budget cut-backs, and the local network may come across a newly identified actor, such as a band of liquefiable silt (its an earthquake zone), which will not cooperate and will render the potential solutions prohibitively expensive. But, as the empirical work of the actor-network researchers has indicated, a loss of actor-network control can also contribute to a project breakdown. Delays or cost overruns may result from the global network being unable to successfully enrol a powerful actor which persists in resisting the sponsor's hegemony. For example, a rail agency may enter a sustained guerilla war to direct resources away from the river crossing project towards a light rail project in a powerful Minister's province. This is a failure of control by the obligatory passage point of the global network. The steel bridge design consultant may start to reverse its own enrolment if, in the process of the feasibility study, it found that all the concrete solutions will be cheaper. It may then approach the Ministry of Steel Production without the knowledge of the project manager to lobby for steel subsidies. This disturbs the concrete bridge consultants and the resulting acrimony may cause massive operational difficulties and a poor result.

Stability and longevity

Having overcome the resistance of the network actors by the process of translation, how is the network stabilized to the degree that the centre of translation becomes unchallengeable? How is the actor-network given longevity? The key lies in the durability of the links in the network. If interaction is only between human beings using speech we must expect the actor-network to be fragile and short-lived. Paper documents add more durability and machines or structures even more. Thus, as time goes on, the material-actors created, constructed and distributed by the centre of translation serve to hold the actor-network together. Thus, feasibility studies, reports, designs and prototypes serve to create, preserve and increase commitment to the network by their role as material transformations of interests and interactions of the actors. Similarly, the durability of communications is enhanced using devices such as meeting minutes, memos and telephone logs. These, together with letters and reports, are often used as control devices where the space between actors represents a threat to the integrity of the network—highly mobile actors sent out from the centre of translation to maintain order.

When a network is dealing with social policy questions, and the centre of ordering is a government body, it is appropriate that stabilization is formalized early. This is often done by creating a steering committee with members drawn from the principal network actors and chaired by a representative from the centre of ordering. Occasionally these committees have gone on to appoint a project manager or arbitrator to carry the project forward (Glasbergen, 1995).

Actor-network research also suggests that the robustness of the network is increased with size. Any increase in the number of fully enrolled members reduces its vulnerability to defection and betrayal by any one member. Like any other network the robustness increases with the redundancy of the network interactions (uncooperative actors can be abandoned and new ones recruited) provided, of course, that the hegemony of the passage point is maintained.

Change itself adds to robustness. In the creation of engineering technology the local network, made up designers and managers, becomes transformed in stages to drawings, specifications and complex objects. As this progress occurs, the global network changes—often abruptly—as the initiators give way to the operations and maintenance actors who use and nurture the transformed local network.

Translation and strategic decisions

We have learnt, so far, that four *network ordering tactics* are of importance if a network is to achieve a successful decision outcome.

1. An effective outcome will depend on a set of manoeuvres called translation, to realign other actors' interests, strategies or problems towards your own.
2. This enrolment into your actor-network may be deemed successful when you are accepted by all members as the obligatory passage point.
3. As an obligatory passage point, you will control all network interactions in both the local network and between the local and global networks.
4. The stability of the actor-network will be increased as the actors' interests are embodied in material objects and with any increase in the number of fully enrolled actors.

In the spirit of Machiavelli, these are all tactics of effective ordering observed through the empirical study of science and technology. No explicit moral principles have been used to judge their social value. However, this is not a sustainable view if we are to arrive at recommendations for good decision making. We must at some time move beyond mere procedure if we are to leap from descriptions of effective manoeuvres to prescriptions for future behaviour.

Modest conclusions to phase I

When I sat down to write this chapter it quickly became clear to me that it was easy to recall events where translation could be detected, but very few cases where the translation process culminated in a successful outcome. How often we see that centres of translation form prematurely and are replaced, and how often they fail to successfully control the key actors—how often, in practice, a passage point fails in its bid to become obligatory. This is the nature of organizing—it consists of an ongoing struggle to achieve the ideal translation. Only rarely is the ideal achieved and we must, in practice, learn to live with iterations towards an approximation. But, in these struggles, decisions are made and action is taken and something, however flawed, is achieved. Not everyone's interests are transformed successfully and some actors will be less than satisfied with the outcome. Because it is a struggle, there will be victims and, sometimes, unexpected winners. Successful translation is also one manifestation of power. Power is an effect of the successful stabilization of a network, manifest in, used and stored by, an obligatory passage point. Unsuccessful translation is a flawed power, a power thwarted by other actor-networks, a power successfully resisted, or a power betrayed. Translation is a process of calculation, negotiation, persuasion, trading and, sometimes, trickery and coercion. In other words, it requires the full portfolio of human skills for gaining compliance from others. As gaining compliance from one other actor is difficult, the achievement of translation with multiple actors is a skill that is rarely wholly perfected and, certainly, a trick that is difficult to sustain over time. Sooner or later everything turns to ashes. But this does not mean that we should turn our eyes away from the messy reality of translation and take refuge in some dream of rational process. We should be content with a suitable modesty and seek only improvement on a human scale.

In terms of the power model we should note that, in many cases, the problem definition is inseparable from the problem solution, and the process of judgement and argument formulation may be short-circuited. In human terms this would be a radical stunting of the search-inference process so important to good decision making. This is particularly true of strategic decision making. If, in the translation process, interests are easily aligned or problem definitions readily accepted then the crucial process of argument formulation is skipped and options remain unexplored. It would seem important, therefore, in any conception of a good decision making process, to avoid premature closure around a problem definition that dictates only a narrow range of options for action.

Also, as stability and robustness of the actor-network is desirable, some attempt should also be made to be as inclusive as possible. We can modestly conclude, therefore, that the following principles (associated with translating the problem) should be adopted if the decision making process is to be deemed effective.

1. We should avoid premature closure around a problem definition which implies only a narrow range of options.
2. The interests of the most powerful actors should be aligned by an agreed problem definition.

These principles imply that you should resist any urge to fix the solution by prematurely defining the problem in a way that appears to be in your interest. To do this is to be naive. For the network to stabilize without undue resistance it is important not to set up unnecessary points of resistance—do not make enemies if you can possibly avoid it. To be sure that you know the limits of the problematic situation and, therefore, all potential problem definitions, it is wise to map the terrain. This is best done by identifying all interested actors as soon as possible and working out their potential interests/problems. I hesitate to specify how this may be achieved as it is the gift of effective leadership that means must fit circumstances. However, a number of formal techniques to aid problem definition may be considered when the decision is of such complexity and importance that they may be justified. Some suitable methods are discussed in chapter 12.

Moreover, whatever machiavellian manoeuvres take place, the end run of gaining compliance will usually be in the form of persuasive arguments. Goal congruity must be perceived by enrolled members of the actor-network. All must be persuaded that the centre of ordering is both inevitable and indispensable. Negotiation, threats and payments may be part of the initial tactics but none of these will be effective without persuasion. And among the paraphernalia of the persuasive arts, a good argument will stand out as the ultimate deal clincher. But, translation will often not be wholly successful—resistance will persist, some actors will remain unenrolled. For the resistors who do not choose to drop out of the actor-network, the struggle will go on to obtain recognition of other potential passage points, of other potential solutions, and other interests to be served. Most of this dialogue will be in the form of written or verbal arguments. Argument is, therefore, the medium of translation and the medium of resistance to translation.

We will now consider how argument is used in the second phase of the decision process.

Phase II: Translating the solution

Having stabilized the actor-network and arrived at a problem definition, argument is used in two ways to achieve the solution. As the power model indicates, the judgements of the human actors concerning the situation and the possible solutions will be finally expressed as arguments, and argument will guide the formulation of the solution such that it will be socially acceptable.

Judgement and argument

You may recall that we do not form our judgements from a direct and comprehensive knowledge of the attributes of the object being judged. Rather, we select certain aspects of the object and base our judgement on those. These aspects, or cues, may be read off the object from a multitude of potential cues or sometimes, are, attributed to the object. Engineers tend to see technical cues, planners see social cues and senior civil servants see political cues, all given the same data in the same circumstances. Moreover, even if two engineers see the same technical cues they may interpret them in very different ways and come to quite different judgements. Well, of course, this model of judgement is roughly equivalent to the results of different arguments concerning the same subject but coming from different fields (Toulmin, 1969; Toulmin *et al.*, 1979). The field of engineering will tend to produce arguments using technical grounds and reasoning and (for example) the field of economics will use very different arguments based on financial criteria. So, as judgements will tend to lose credence with people from other walks of life, so arguments lose force when they cross into other fields. In argument theory terms, the cues are the 'grounds' and the judgement is the 'claim'. Thus, as the same cues can produce different judgements, so the same grounds can lead to different claims. It is but a short step, therefore, from social judgement to social argument, and in the process the apparently positive judgement process becomes infused with purpose. The goals that were implicit in the inference process are explicitly revealed, and in so doing are made political. What was cognitive and personal has become part of human dialogue—part of the social complex, with all its message-bearing advantages and disadvantages. The judgement is now part of language and as such has itself become a set of cues for interpretation. The argument/judgement/cue will now carry a value-soaked significance such that the words, however plain and straightforward, will inevitably convey slightly different meanings to different actors (Eiser, 1990). Nevertheless, language is our most comprehensive and retraceable interactive mode, and argument is its most powerful medium of persuasion.

Argument as an expression of judgement.

As participants in the actor-network conduct their dialogue in the translation process, the data for judgement will be increasing rapidly. The engineers will be doing calculations, planners will be considering organizational and social implications, accountants the costs, and the relevant technology and environment will either be cooperating or resisting. Often the data is generated by a search process and sometimes by attempts to design alternative solutions to fit the problem definition. Nutt (1993b) has demonstrated that solution alternatives may often be sought from suppliers or outside contractors or be modifications of other successful solutions. Thus, the data may not be totally disaggregated but may take the form of partially combined solution units. However it is generated, data will be revealed, hidden and manipul-ated during the manoeuvres of translation but, ultimately, most, if not all of it, will be used as grounds or backing for arguments for or against solutions. As we have noted earlier, there will be a tendency for particular roles to utilize role-relevant cues in their judgements and these will mould the contents of their arguments. As the arguments defuse throughout the network, by formal and informal means, alliances will form and complex, multimodal arguments may develop for a particular solution. This is assuming that premature closure has not occurred such that the problem definition will only allow one solution. This, of course, is a radical short circuit of our model and, because it truncates the data available for decision making, it must be considered undesirable. Good decision processes should attempt to enrol the actors into a definition of the problem such that a range of solutions are possible. With a common goal, multiple potential solutions, and an open dialogue, the judgement will enter the network as persuasive arguments. Without one of these conditions, conflict and distorted communi-cation will be the norm.

Argument as a measure of value fit.
Sooner or later the final centre of ordering will desire to choose a solution. This centre may not be identical to the agency that initiated the process because the process of expanding the network may have triggered a realignment towards another more effective actor. And, furthermore, as time goes by, and interests become transformed into artefacts, the actors change and the centre of ordering may shift. For example, a computer firm may suggest the need for an organization to install new data processing equipment. This may be accepted and a centre of ordering (obligation passage points) may form around the project manager. Dialogue about the type and scale of change may be free and well informed. Reports will be made and memos exchanged. The project manager will make recommendations in the form of a well argued report to the managing director. But, the relevant centre of ordering has now shifted from the local to the global network, from the project manager to the managing director. It is the final centre of ordering, the managing director, who will make the ultimate choice. How is this choice made?

It is my contention that the decision maker will make the choice using the following logic

1. Does the short list of potential solutions fulfil the minimum requirements to achieve the original goals? (This fits the satisficing, incrementalist, conservative pattern of most organizational decision making.) If the answer is yes
2. Which solution will require the least power to implement? Again, this implies that the urge to optimize is not a factor in the final choice but, rather, a concern with the long term viability of the decision. Will it stick?

Before we go on, the type of power referred to in point 2 (above) needs some explanation.

Franco Crespi (1992) describes one function of power in the following way. People create the actor-networks we choose to call social systems. These are most obviously experienced by the individuals as social institutions which regulate the dynamics of the national, regional, local area and functional entities. These institutions are products of social ordering and encode and make concrete the shared meanings, values and norms of our group experience. In a similar way, written and unwritten codes of conduct regulate our individual behaviour in and around these institutions. These institutions and norms are highly simplified guides to action and, if we relied on them totally, social life would stagnate and the culture would collapse. In any particular situation, the institutional/normative actor-network will often not be suitably configured to fully guide our individual actions or inform our judgements. The decision requirements of life are far too complex for us to conceive of a set of rules to cover other than a very few situations. Those local actor-networks (individual and collective), who are required to take action, live in a region of tension between the desire and necessity for unfettered action and the experience of the constraints of the greater social ordering. The effect generated by the local network to cope with this contradiction, and enable it to make particular decisions and take local action is what Crespi (1992) calls 'outer power'. This power enables actors to legitimately interpret norms or rules, to adapt them to a particular situation, to fill in the normative voids if required, or to bend and adapt the social order to the requirements of the action. The greater the need for interpretation, innovation, or adaptation of the social system and its rules to the decision/action needs of a local network, the greater will be the power required. Power, then, is the effect of the process of successfully managing the contradictions between the complex needs of decision making and action, and the constraining simplicity of the surrounding social order. This, then, is the key to the final role of argument.

Resistance to the solution, developed during the proceeding stages of the process, may be so strong and persistent that the centre of ordering may not proceed with implementation. Strong resistance is often based on values not salient to the centres of ordering and, therefore, their effect is underestimated. Environmental values are of that nature when expressed in an engineering context. However, the momentum of decision making, the embodiment of past decisions in reports, files and public commitments, often tempt the centre of ordering to push ahead. The result is often the triumph of power in the short term and the bitterness of failure in the long. Those wise enough to hold off will often return the problem for redefinition, in the hope that all that had been learnt and lost will eventually help the formulation of a problem/solution that both addresses the problematic situation in a sufficient manner and can be sustainably implemented.

General conclusions

Experience indicates that many powerful actors will attempt to quickly define a problem in such a way that only a very limited range of solutions are possible. This is akin to aborting the search process. Over time, if this decision process is repeated, poor decisions will result. Effectiveness will be the result of luck rather than a robust and effective procedure.

What this chapter has attempted to illustrate is the way persuasive argument precludes premature closure around a particular problem/solution. Argument also has a profound ability to express human judgement, such that the covert nature of judgement is transformed into an overt and purposeful structure (grammar) for comparison with judgements arising from different cue selectors. Care is needed, however, because the vocabulary of each argument is still role-dependent and may not convey meaning to an observer from another field of endeavour. Indeed, one of the gifts of a skilled manager is to remain open to varieties of vocabulary and, if necessary, to act as a trusted interpreter within the dialogue space.

But, in terms of the ability to make decisions that will have a good chance of survival over the years, argument has an even more important role. Arguments cannot be constructed without embodying the value system that influenced the cue selection and inference process of the relevant actor judgement. Similarly, these values imbue the equivalent structure of argument with a coherent significance which enables its social meaning to be seen. Thus, arguments can be used to measure the fit between a judgement and various value systems. This enables some decisions to gain the robustness of the wide social acceptance so desirable in an organizational or social setting. If a less than ideal fit occurs between the solution and the dominant normative structure(s), a secondary process may be set in motion to reinterpret the norm, invent a useful substitute or bend or adapt it in some way. These manoeuvres require the resources, freedom and purposefulness of action which will be seen as power—power to adapt the social to the requirements of decision and action.

The lack of fit is a manifestation of resistance within the actor-network to the proposed solution based on incompatible values. Although the power required to overcome this resistance and improve the fit may not be overtly coercive it will no doubt be experienced as such by individuals. This will often result in driving the resistance to a deeper level. The momentum produced by the transformation of interests to artefacts and the robust stability of a wide network may well sustain the actor-network despite deep resentments. Nevertheless, given the opportunity, the dissatisfied parties will bite back.

Experience also indicates that the generation and expression of arguments based on roles can produce surprises. Certainly, opposing arguments tend to surprise professionals such as engineers. A positivistic way of life centred around the manipulation of matter is not the best preparation for social dialogue. Engineers also tend to underestimate the fragility of technical arguments. Every judge can attest to the ability of competing parties in a court case to undermine the credibility of expert witness using opposing arguments from the same field of endeavour. The facts seem so much less certain when challenged from within. Social arguments are less quantifiable but nevertheless, are, more robust, because the value-laden nature of the grounds and backing is more easily recognized. Nevertheless, all counter arguments tend to come as a surprise. They disturb our role-centred life and reveal the richness and diversity of judgement. This is the reason why the opportunity for argument is so important to an effective and long lasting decision. Arguments add to our own limited store of data and hand us the tools for action.

We may conclude, therefore, that this chapter indicates four network ordering tactics and five decision principles that are generally needed to achieve a good strategic decision.

Four network ordering tactics

1. An effective outcome will depend on a set of manoeuvres, called translation, to realign other actors' interests, strategies or problems towards your own.
2. This enrolment into your actor-network may be deemed successful when you are accepted by all members as the obligatory passage point.
3. As an obligatory passage point, you will control all network interactions in both the local network and between the local and global networks.
4. The stability of the actor-network will be increased as the actors' interests are embodied in material objects and with any increase in the number of fully enrolled actors.

Five decision principles

1. We should avoid premature closure around a problem definition which implies only a narrow range of options.
2. The interests of the most powerful actors should be aligned by an agreed problem definition.
3. The decision process should encourage the formulation of a wide range of sound judgemental arguments.
4. The final solution should be based on those arguments that point to a good fit between the solution and the dominant receptor value system.
5. Should the fit be incomplete, and excessive power is required to adapt the social network to the requirement for action, then consideration should be given to a reformulation of the problem.

Strategic decision management—an illustrative case study. 'JVP Engineers'

JVP Engineers performed consulting assignments for the power industry in Europe. They had grown to a staff of about 500 over 20 years and were well respected in the industry as a reliable, if somewhat conservative, performer. The sector specialization required close contact between the principals and key client representatives. They were well aware of the parochial sensitivities of the national power authorities and carefully promoted the localness of their offices.

In the London head office the chairman and managing director were dissatisfied with the overall performance of the firm. Profit after tax had been less than 3% for two years in a row. As the power industry in Britain was entering a stagnant phase, it was obvious the answer to increased profitability lay with the two larger regional offices in Rotterdam and Paris and the two small offices in Edinburgh and Brussels. Structural problems were evident however. Each office prided itself on providing the full range of services to the clients which resulted in some disciplines being represented by a single consultant. Economies of scale could not be achieved if the full service requirements of the clients were to be satisfied.

The relationship between the chairman and the managing director was cordial but a little distant. The chairman, who had founded the firm 20 years before, had just promoted the chief engineer to managing director two years ago despite some misgivings about his entrepreneurial skills. However, on this issue they were agreed. The operations manager would be assigned to visit each regional office, talk with the managers, staff, and clients and report back to head office. The operations manager had been with the firm for ten years and was close to both the chairman and the managing director—although some lingering resentment about the promotion of the chief engineer could be detected when he was into his third beer. In fact, he felt that it was he, rather than the managing director, that ran the firm. He controlled the accounting and recruitment and, in this case, it was his report to the principals that had precipitated action. He was aware that the regional managers were not always happy with his power to influence outcomes by administrative or persuasive means, but he was confident that his grasp of the big picture was appreciated by all.

A meeting was called to set the scene and sell the operations manager's assignment. It was clear from the start that the Rotterdam and Paris managers attributed the lack of profitability in their regions to an excessive head office overhead being loaded on to their books. They were working to capacity and had reasonable charge rates but overheads were soaking up the profits when the accounts were presented. A cut in head office overhead was required—particularly in operations. The managers from Edinburgh and Brussels carefully kept their heads down. The managing director said that blaming the head office overhead was ducking the issue. He thought the problem perhaps lay with the lack of marketing, resulting in a suboptimal size of each office, or perhaps they should target the market more and forego breath? The issues remained unresolved and the operations manager was assigned to review the regional offices.

The operations manager felt his tour of the offices had been a success. He had examined the accounts, staffing levels, client lists, current work, forward work prospects and had informal chats with client representatives. He reported that the clients appreciated the full service profile of the firm and recognized that charge rates had been set to reflect this lack of specialization. Staff morale was also linked to pride in their ability to tackle any type of problem. Managers were flat out trying to juggle the requirements of proposal writing and consulting. It was clear to him that regional profitability could be enhanced by more support from head office to allow regional managers to concentrate more time on consulting. He recommended that all accounting be centralized (including client billing) and key proposals should also be the responsibility of head office. He congratulated himself that the two minor managers were on his side and that the managing director would be keen to use his considerable technical expertise in proposal writing.

The plan to centralize the accounting and billing was vigorously opposed by the two larger regional offices for four reasons. Firstly, it would undermine the independence and power of decision making of the regional manager; secondly, it would slow the billing process; thirdly, it would give the local power authorities an impression that they were not native, and finally, it would slow payments to subconsultants and alienate them. They were particularly concerned that technical decision making was already complicated by a matrix structure dominated by chiefs in head office. Loss of financial control would make them mere satellites and morale would disappear. Proposal writing support was welcomed, however, but did little to defuse the anger. One of the regional managers, who was at that time controlling a very large assignment, hinted that she might resign if this plan was adopted. A major row was brewing and the next coordination meeting was quietly cancelled. The chairman flew to meet with the two major managers and agreed to start again on the problem. This time the question of head office overheads would be addressed in two ways. For the smaller offices, the operations manager's plan would be used. For the larger offices, they would be treated as independent firms and would control all regional-specific operational and accounting functions. This deal effectively cut out the managing director and as a result trust between the managing director and chairman was irretrievably lost. The operations manager felt relieved to have retained his skin but was unrepentant. As a result the scheme worked in the short term but was eventually undermined by the combined resources of the managing director and operations manager.

Case study discussion
How can this decision process be described in terms of the power model?

1. *Constraints.* The first situational constraint to the problem definition and potential solutions resolved around the geographic isolation of the branch offices and the parochialism of the regions. It was clear that a local full service was valuable to clients. Secondly, the intermediate size of the firm precluded complete independence for each regional office. Threats to egos and entrenched interests—particularly those of the operations manager, permeated the story, spiced by the suppressed ambivalence of the chairman to his new managing director.
2. *Problem definition and enrolment.* Although the problematic situation was recognized by all the principals, it was the managers of the Rotterdam and Paris offices who first defined the problem. However, this was premature as key figures had not been enrolled into an alliance. The operations manager had both a strong interest in retaining control over the region's accounting and the ability and opportunity to enrol the chairman and managing director. He then set about putting in place a procedure which would allow him to act as the obligatory passage point and furthermore to speak on behalf of the clients and staff. Thus, the problem and the logic of the solution were his.
3. *Communication, argument and resistance.* As the acrimonious

arguments developed the two managers of the smaller regional offices wisely kept quiet as they feared that they could be damaged in the exchange. Communication between the parties took place in the formal settings of meetings and informally by telephone and personal contact as each manoeuvred for allies. Resistance was fierce and a meeting was cancelled to avoid overt confrontation.

4. *Judgemental arguments.* The arguments first spoken by the managing director concerning market focus were made irrelevant by the operations manager's problem definition. This definition and the proposed solutions were, in effect, merely incremental extensions of the current situation and ignored contrary data advanced by the regional managers. The arguments of the regional managers were soundly based in the practicalities of running a regional office.
5. *Final judgement formulation.* It was clear to the chairman that the cost of imposing the operations manager's solution on the entrepreneurial and managerial values of the two larger regional offices would be excessive. Wisely, he sought a redefinition of the problem and a compromise solution that would fit the marketing and organizational values of offices of both sizes. The resulting damage to his relationship with the managing director could be accepted as it reinforced his previous doubts concerning the managing director's judgement. In the long term the decision did not stick, as the long term bitterness of the managing director and the operations manager, due to the breach of fundamental organizational discipline, was sufficient to constrain its effectiveness.

A better way?

What would have been a good decision process given this problematic situation? Alternative procedures are suggested by the five principles stated earlier.

1. *We should avoid premature closure around a problem definition which implies only a narrow range of options.* It would have been

wise for the chairman to make sure that an obligatory passage point did not form too early. This could have been done by defining the problem in such a general way that the definition would not point to a solution. Premature closure around a definition and solution is sure to alienate key actors. Perhaps an emphasis on the firm's profitability at the first meeting and an invitation to the others to provide potential options for action would have prevented closure.

2. *The interests of the most powerful actors should be aligned by an agreed problem definition.* Had the chairman chosen 'falling profitability' (or similar) as the problem, there is no doubt that all the key actors would have agreed on its importance and would not have prematurely resisted the decision making process.
3. *The decision process should encourage the formulation of a wide range of sound judgemental arguments.* Perhaps the conclusion of the first meeting should have been the formulation of a range of options with a request that each actor should draw up a list of good reasons, backed up by data, for and against each option to be circulated to all committee members. Each player in the action should be warned that arguments for or against solutions will be tested against a thorough analysis of each regional office's financial, technical and market position. This data would be gathered jointly by the operations manager and the regional manager. Arguments should also acknowledge that consultancy is about people and their interests, and this should be reflected in the various position papers.
4. *The final solution should be based on those arguments that point to a good fit between the solution and the dominant receptor value system.* In this case, perhaps the chairman or managing director could have formulated the solutions with the strongest consensual arguments and circulated them for comment and resolution at the next committee meeting. Should anyone's rights or interests be violated, the value of fairness could have been invoked to enable suitable organiza-

tional compensation to be devised. Such an open process would have the practical bonus of producing long term commitment to the solution from all who were part of the decision.

Summary

- This chapter has used the power model to illustrate how strategic decisions may be managed.
- The first task is to translate the problem such that all interests are aligned and the key players are enrolled into your network.
- The second manoeuvre involves the use of argument to express actor judgements, and to resolve conflict between arguments in such a way that the chosen solution does not require an excessive use of power for its social acceptance.

12

Tools for strategic decision management

People are stimulated to make decisions by problems or opportunities. Furthermore, the decision process is determined, to a large degree, by the assumptions surrounding the definitions of these opportunities or problems. Often, the clear cut nature of many engineering problems produces well defined and retraceable decision processes. However, when an engineer moves away from design into the risky world of organizing, the decision processes tend to be less well modelled. This is partly due to the non-analytical nature of management work, which deprives managers of an agreed knowledge base and the comfort of scientific rules. It is also a function of the ambiguous nature of management problems and solutions—if the problematic situation remains fuzzy then the decision strategies will be less than obvious. It is the purpose of this chapter to describe some methodologies that can help a manager reduce the misunderstandings that may arise from inadequate modelling. We will start with the analysis of the interests and cues likely to be involved and move on to some well known techniques used to flush out ideas from a group. Such moves are important because they put the manager in touch with a range of ideas and improve the chances of final group acceptance of both the definition and the solution. Finally, the structure of the persuasive arguments required to express our support for potential solutions and test their fit with the dominant value system are discussed.

Analysis of interests

When a manager becomes aware of a problematic situation, or recognizes that an opportunistic environment exists, the most sensible initial action he or she should take is to convene a closed door meeting of trusted compatriots to identify the potential actors and their interests. To be done successfully, a familiarity with similar networks is important—particularly if other powerful organizations or government departments are likely to be involved. The following steps are required

1. loosely define your interests in relation to the situation
2. list all potential members of the network
3. loosely define the likely interests of each network member in relation to the situation
4. describe the positive and negative relationships between your interests and those of the other actors
5. rank these relationships in terms of relative power to effect positive outcomes
6. devise a number of problem or opportunity definitions that take account of the most powerful interests without violating your own.

If this process indicates that the most powerful interests can be successfully aligned with yours, proceed to more closely define the decision process.

Anticipating cue sets

We have learnt from chapter 7 that each individual is likely to perceive a situation slightly differently. Different cues may be selected and different weights placed on those cues. We know that cue selection is heavily determined by roles. With this in mind, we can anticipate how each actor in the network will model the problematic situation or opportunity. The engineers are likely to use technical factors in their judgement of the problem but public servants may see a social problem. Thus, using a knowledge of roles and interests you can devise a definition of the problem/ opportunity that is sufficiently inclusive of the principle cue sets to enable the network enrolment of the most powerful actors. The exercise will also enable you to anticipate the likely content of each of the actor's arguments concerning possible solutions.

Brainstorming

This technique is designed to generate as wide a range of problem, opportunity, or solution attributes as possible. First of all, the problematic situation is stated as simply as possible and recorded in the centre of a white board. A statement like 'production is falling' is adequate. Members of the group are then asked for reasons why they think this situation exists. All answers are acceptable except those couched in terms of rigid solutions. The fringes of the problem should be explored in a non-judgemental fashion.

No criticism is allowed and as many ideas as possible are recorded on the white board. This process may be sufficient to reveal agreement about the problem boundaries. Often, however, a further degree of analysis may be necessary. One useful way to model the answers is to place them into categories in a fishbone pattern. Six fishbone headings have been suggested by Robson (1993) (people, environment, methods, plant, equipment, and materials) but others may be more suitable for networks. Fig. 17 shows a typical fishbone for people.

The other ideas thrown up by the brainstorming session can be similarly modelled under the other category headings—all pointing towards, and serving to define, the brief statement of the problematic situation.

Nominal group technique

This is quite similar to brainstorming but somewhat more controlled (Delbecq and Gustofson, 1975). Each group member is asked

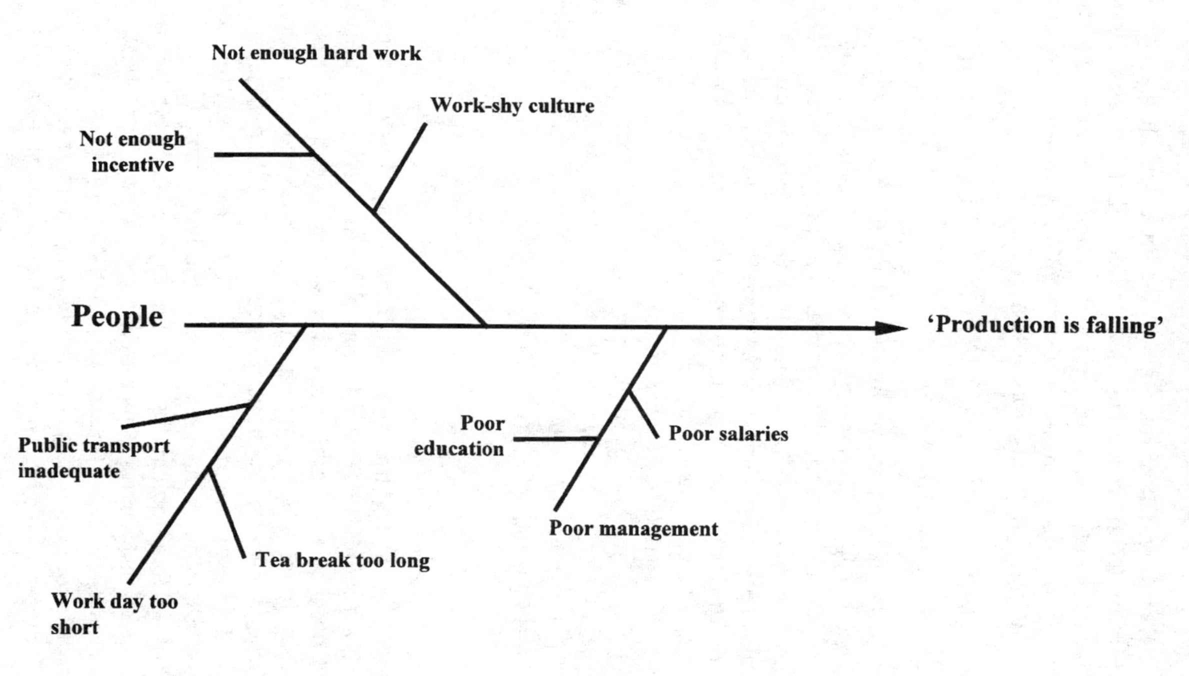

Fig. 17. Typical fishbone diagram

to create their own list of reasons for the situation. They are then asked, in turn, to nominate one reason which is placed on a white board master list in a circle around the statement of the problematic situation. These ideas may then be ranked by the individuals for subsequent aggregation and discussion.

Delphi

This method does not require the group to meet. Each group member is asked to give reasons separately which are then collated and recirculated. With this list in mind, each is again asked for a revised set of reasons. The technique was developed by RAND Corporation for technological forecasting and is particularly useful for the production of convergent numerical estimates (Armstrong, 1985). As an idea generator it is probably not as useful as the previously discussed techniques.

Often situations are of such complexity that there exists no substitute for an in-depth, structured interview with each important player. A means of directing and modelling such interviews has been suggested by Eden *et al.* (1983).

Cognitive maps

The central concept of this method is the *personal construct* derived from the work of Kelly (1955). A personal construct is a picture of a concept described in terms of its positive and negative attributes. Thus, the phrase 'not enough incentive' can only be fully understood when viewed in the light of its conceptual alternative. In this case it could be 'just enough incentive'. The personal construct then becomes 'not enough incentive ... just enough incentive'.

First of all, the problematic situation is given a label in the same way as in brainstorming. This may be 'production is falling'. This construct is clarified by thinking of a satisfactory alternative such as 'steady production'. Placing this construct in the centre of a sheet, the interviewee is now asked a series of questions about *why* they are concerned about the problem—again, expressed as constructs. These are then connected by arrows with positive or negative signs to indicate causality. The lower half of Fig. 18 illustrates this stage.

The second stage starts further out and works towards the label, noting constructs that map the *reasons* for the problem. This is illustrated by the upper half of Fig. 18. Further questioning can also use 'why does it matter' and 'why is it like that' to elaborate on some of the secondary but important constructs in the map.

Naturally this process is time consuming and highly skilled—particularly when a number of interested parties have to be interviewed. The resulting cognitive maps are, however, powerful and complex definitions of the problematic situation and a potent stimulus for the negotiation of an agreed strategy.

In strategic decisions that involve social policy, two important modelling techniques have been used that attempt to define problems and solutions in terms of their constraining and enabling attributes.

Value trees

These are sometimes called goal hierarchies and conceptually link high level values or goals to practical descriptors. They are one stage of an analytical technique in decision theory and, therefore, are described in more detail in chapter 6. Fig. 19 illustrates a value tree used to model the features of a question concerning the evaluation of road network links. It starts from a very high level value or goal and describes how the attributes of an ideal road network would be linked. Although insufficient data may be available to enable the model to be used in its full decision analytical mode, it usefully points to the constraints surrounding the problem space.

Scenarios

Creating imagined futures is another useful way to explore the boundaries of the decision space and link problems with potential

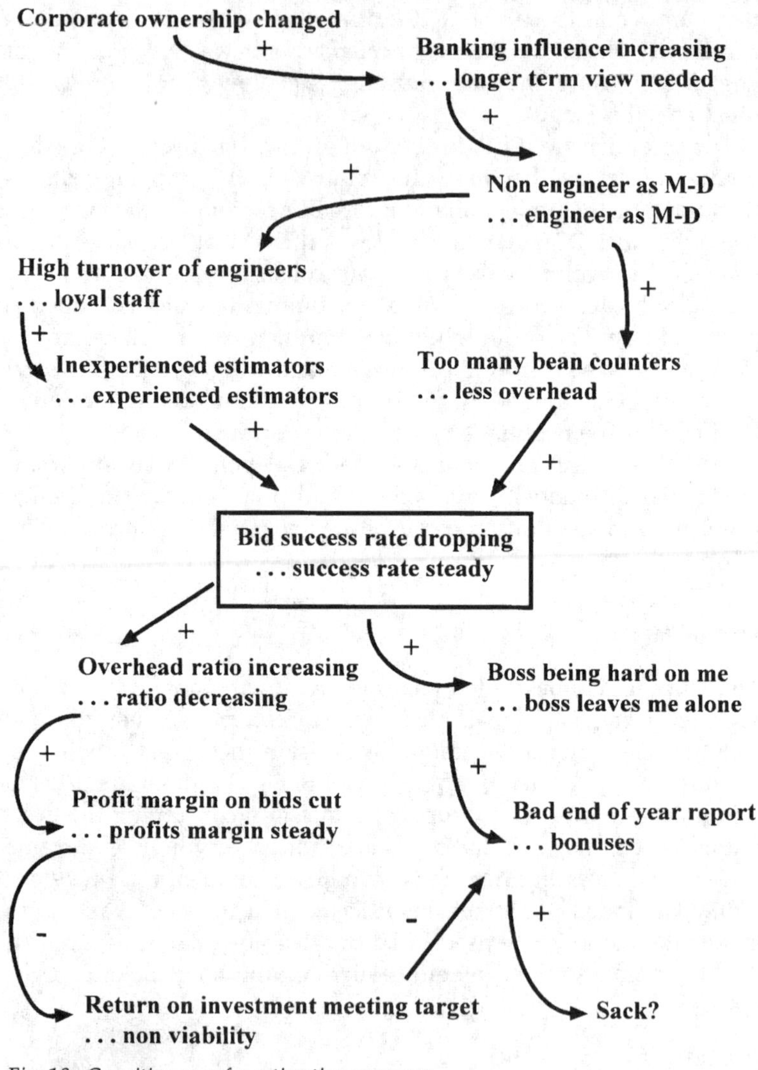

Fig. 18. Cognitive map for estimating manager

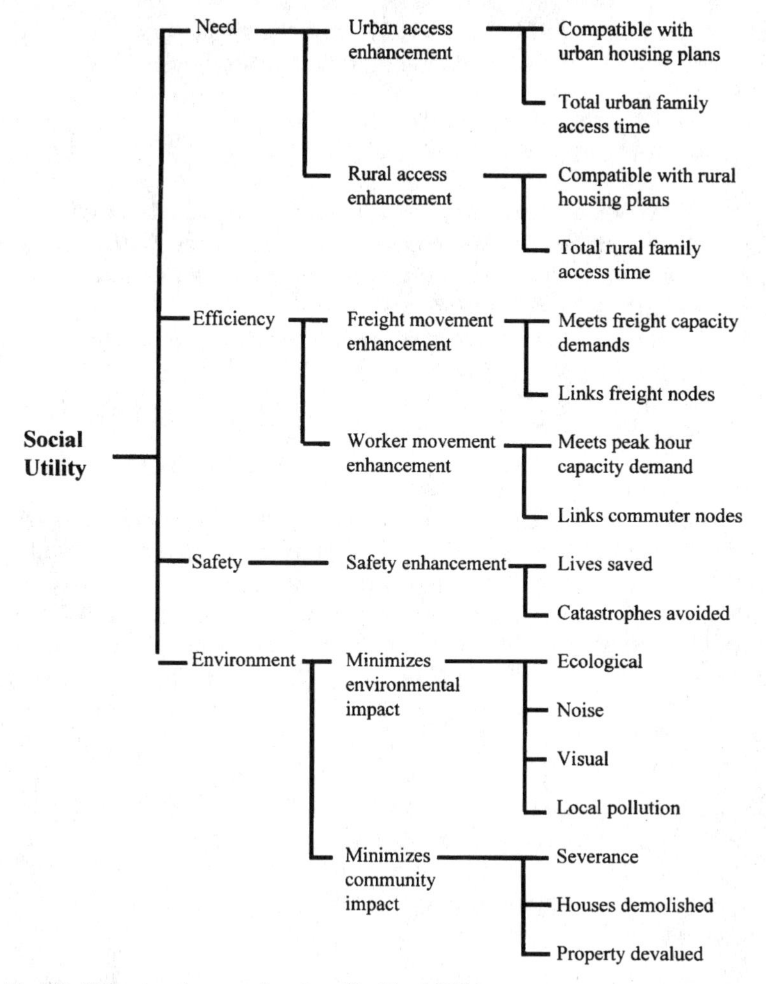

Fig. 19. Value tree for a road system (Parkin, 1994b)

solutions. Ducot and Lubben (1980) describe six factors which may be used to describe scenario generation approaches.

1. *Exploratory.* These scenarios start from the present and explore the consequences of different sequences of future events.
2. *Anticipatory.* A snapshot of a potential future is described and the events needed to achieve that scenario are deduced.
3. *Descriptive.* These scenarios are created without any judgement as to their desirability.
4. *Normative.* Normative scenarios are created on the basis of social values and desired goals.
5. *Trend.* These scenarios are based on forward projections of past data.
6. *Peripheral.* These scenarios explore low probability futures.

Engineers and managers like to use *exploratory–descriptive–trend scenarios.* Typically, future transportation infrastructure requirements are explored using high and low projections of available demographic and traffic data. Judgements of desirability are often restricted to technical questions, such as congestion. Citizen groups, however, tend to use *anticipatory–normative–peripheral* methods when considering possible futures. They often like to describe ideal futures based on particular value systems and explore what policy measures would be required to achieve them. Unfortunately, these ideal scenarios tend to be considerably less likely to occur than those produced by trend analysis.

Schoemaker (1991) has recommended a number of steps to create limiting scenarios that can be used for exploratory or learning purposes.

1. decide what you wish to understand and review its history
2. identify stakeholders, interests and power concentrations
3. identify and quantify trends
4. identify the relevant uncertainties and their relations
5. construct two forced scenarios—one using all positive outcomes and the other all negative outcomes

6. eliminate all combinations that are internally inconsistent in terms of trends and outcomes, and reconstruct
7. estimate the future behaviour of the stakeholders within each reconstructed scenario
8. if possible use techniques such as Monte Carlo simulation to examine the key uncertainties
9. after re-examination of these scenarios, deduce decision scenarios from each.

The creation of scenarios is a useful means of anticipating the future behaviour of the network actors, given a knowledge of their interests and probable cue sets. They can also indicate some of the things that may go wrong in the management of the decision process.

Information Technology (IT)

In some situations, where numerical data is critical, a limited view of the decision constraints may be obtained from the IT data banks available to large organizations.

IT is the computer technology for data storage and processing. IT systems may contain only an efficient means of accessing and structuring large bodies of data or contain within them some means of analysis and simulation. If we assume that more information enhances the number and quality of cues for judgement, we must also assume that IT systems will enhance the decision process. Recent work by Molloy and Schwenk (1995) appears to confirm that this is so. After examining a number of real decisions they concluded that, provided the decision process was not performed in a crisis environment, the use of IT produced at least six improvements, with the most important being the first

1. the identification of problems and opportunities was more rapid and more accurate
2. more cues were available for judgement
3. more alternative solutions could be examined

4. network communication may be enhanced
5. higher quality decisions were achieved
6. the decision making time was shortened.

However, the authors also point to the dangers associated with an over-reliance on suspect data, and the gross errors that may result from ignorance and lack of practice in the use of a particular IT system. It should also be emphasized that strategic decisions never contain only numerical attributes and computer analysis can only contribute in a very limited way to the process.

So far we have discussed various methods of generating attributes of the problematic situation to enable you to put together a definition that will enable a well managed decision process and suggest solutions. Sometimes gaps and redundancies can be detected by attempting to place this information in a standard project management format. One such format, the logical framework is particularly useful.

Logical framework

This simple methodology was created by the United States Agency for International Development (USAID) to structure the definition of overseas development projects (Coleman, 1987). It requires a four by three matrix of cells to be completed which between them define the project. On one axis are four hierarchical values.

1. *Goal.* The high level objective of the project.
2. *Purpose.* The effect that should be achieved as a result of the project.
3. *Outputs.* The actual results that contribute to the purpose.
4. *Activities.* The activities that have to be undertaken to produce the outputs.

On the other axis are the indicators and constraints.

A. *Verifiable indicators and inputs.* The measures to verify to what extent the goals, purpose and outputs are achieved and the goals and services necessary to undertake the activities.

B. *Means of verification.* The data sources necessary to verify the status of the indicators.
C. *Assumptions.* These are the external events, conditions, or decisions necessary to achieve the goal, purpose, outputs and activities.

For the purposes of problem or opportunity definition and the generation of solutions, the ability to clearly state the goal, purpose and assumptions to be achieved by the decision process will be determined by the comprehensiveness of the data provided by some of the techniques discussed earlier. However, do not let the lack of information stand in the way of creative action.

Good argument

Arguments are at the heart of the power model and their quality is critical to their acceptance. The philosopher Toulmin (1969) has suggested that a sound argument has a structure similar to that shown in Fig. 20. The data (cues) (D) relevant to our argument are shown on the left of the diagram and the aim is to move, via a warrent statement (W) to a claim (C). To be sound, a warrent must have adequate factual backing (B) and an added robustness is created by an admission of doubt (Q) and the discussion of any possible rebuttals (R). Fig. 20 shows a simple argument expressed in only one paragraph. However, more complex arguments, which may be contained in full reports, should contain within them similar elements.

A warning

The tricks of the trade discussed in this chapter are useful for sorting out the complexity of real situations. There is no substitute, however, for dialogue, logic and, above all, experience. A finely tuned political antenna helps as well! We should also bear in mind that many real decisions are made without bothering too

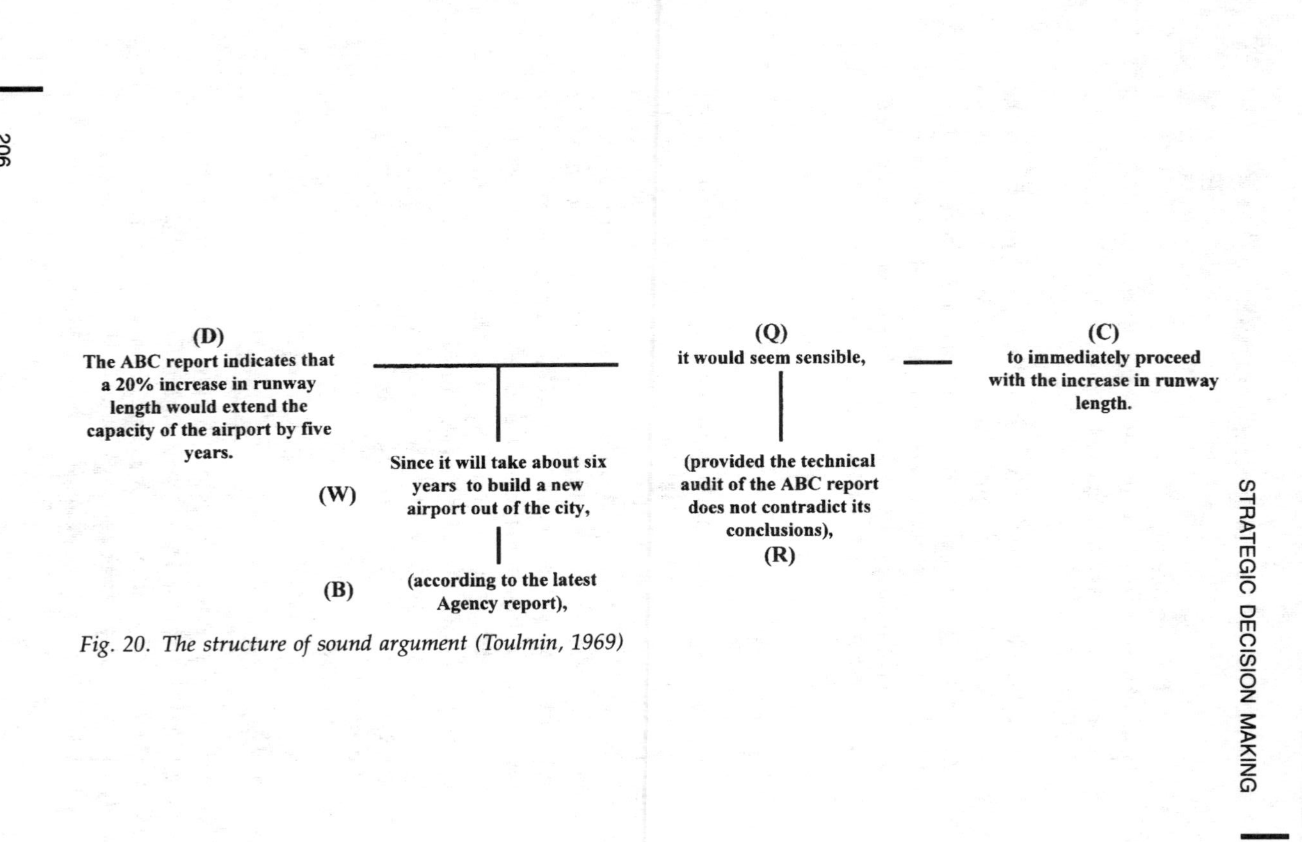

Fig. 20. The structure of sound argument (Toulmin, 1969)

much about a detailed definition. Nutt (1993b) found that, of 336 decision cases studied, only about half generated solutions using a search procedure or solution design driven by an explicit problem definition.

Strategic decision tools—an illustrative case study. 'The new government'

The new State Government has taken power. The Minister has indicated to the Secretary for Transport and Urban Planning that by next month he wants an action plan to address the growing doughnut effect in the city of Metropolis. How is the Government to reverse the steady migration of population and business from the inner city to the suburban rim?

Case study discussion

The Secretary for Transport and Urban Planning would be wise to follow a procedure like the one described below.

1. She convenes a number of meetings attended by senior representatives of key urban departments to identify whose interests will be affected by any change in urban policy. Certainly, government departments concerned with the provision of infrastructure such as roads, water supply, sewage, public transport and housing will be affected. Business interests will also be affected if the land use in the Central Business District (CBD) and inner suburbs is changed. Local councils and community groups are important.
2. As there is no time for consultation with these groups, a follow-up meeting attempts to describe the attributes of the problematic situation which will be of most importance to each of the groups. These are the cues that these groups will use in any judgements they may make concerning the situation. Naturally, the committee would anticipate that technical cues would dominate the infrastructure depart-

ments and green cues the environmental groups. The reactions of the business community will be varied and dependent on present business location, and those of the councils may be difficult to anticipate because of local political factors.

3. Armed with this information, and the contents of previous studies of the problem held in the archives, an aide is assigned to build a value tree from the attributes and anticipate how each of the interest groups would weight the attributes.
4. At this stage the Secretary for Transport and Urban Planning decides that there is a danger that political considerations may be ignored. She decides, therefore, to interview the Minister in depth using the cognitive map technique. The result is that a number of important political attributes are exposed and these are built into the value tree.
5. Finally, a draft brief for consultants is drawn up using the logical framework and placed before the Minister for approval. With information derived from the value tree, the covering letter describes the anticipated reaction of the key interest groups to a range of possible solution types.
6. A consultant is appointed. To ensure an unbiased report, none of the previous work is revealed to the consultant.

Final remarks

Parts II and III of this book divide decision making into analytical and holistic approaches. In my experience, there is a tendency for engineers to overvalue analytical processes and denigrate those of a non-analytical nature. In terms of decision making this would be a mistake. *All good decisions are holistic*—even if the judgement is partly based on important technical cues. I know of no management decision making context where factors of a non-technical nature are not of great importance. To put it bluntly, if you have made a decision based on calculations only, then it is probably inadequate. Nevertheless, if you have the time, by all means utilize your analytical skills and use techniques such as decision analysis —the process may usefully throw up new cues or cluster seemingly disconnected and confusing cues. But in the end, remember that the final judgement must be holistic.

Decision making is a label we place on certain points of resolution in the complicated flux of organizing. It has proved a convenient concept to bring together psychological, sociological and mathematical models of how such solutions are achieved. But, in the words of John Law (1994, p. 43), in any context there are 'endless different modes of organizing ... a network of different worlds'. Thus, the dominant administration and enterprise modes noted in chapter 2 produce a tendency towards a greater or lesser reliance on formal organizational rules to shape decisions. In a similar way, organizations dominated by professionals are more likely to utilize either recognition-primed decisions or analytical methods.

However, we should not be misled by the sheer volume of studies on decision making into attributing this act of resolution with any magic powers. Like notions such as leadership, decision making is important primarily as a significant concept in our understanding of organizing. Indeed, it is an act of faith to assume that good decision making, however that may be defined, is linked to satisfactory organizational outcomes. Although it may be common sense to make such an assumption, we should beware of personalizing the process in the same way that leadership has been. We have all had to suffer the boredom of keynote speakers, usually chief executives, who attribute the good years to their leadership, and either gloss over the years of poor performance or attribute them to particularly unfavourable market forces. Similarly, the image of the top decision maker should not mislead us into placing too much importance on the acts of individuals. Of course, some people will be better at producing positive outcomes and this may well be traced to a superior decision making ability. However, others may be equally effective without any discernible grasp of the principles of decision making. This is because effective organizational outcomes are normally the result of a complex web of interactions involving many people, technologies and institutions, and any one person is just one node in that network. Nevertheless, the concept of leadership and decision making do come together when one considers the role of the obligatory passage point or centre of translation. In many organizational or social decision making processes, an effective outcome is critically dependent on the work of an individual or small group in the translation stage. The act of achieving a satisfactory definition of the problem and the alignment of other actor's interests is leadership in action. Stabilizing a single definition, as the result of power and persuasion, is the achievement of good organizing and the essential precursor to the production of a satisfactory outcome. Therefore, as a general principle, I would recommend all managers to retain the concept of the organization (and its external links) as an actor-network, and to look upon the process of organizational decision making as driven by the effective management of social interactions. This will enhance not only the effectiveness of decision making as a social act but also the ability of you, as an individual, to survive and prosper.

Summary

- If the situation is complex and unclear, some or all of the following techniques may be used to produce a definition of a problem or opportunity and suggest solutions that are both inclusive and tractable.
  1. The first and most important strategy is to attempt to model the situation from the point of view of the most powerful network actors by analysing interests and anticipating cue sets. Brainstorming, nominal group technique and delphi may be usefully employed.
  2. If possible, interview key actors and model the detail of the situation using cognitive maps.

3. Use value trees and scenarios to map the possible attributes and boundaries of the problem/opportunity. These techniques will give you a feel for the logical interconnectedness of the various factors, the likely end goals, time scales and probabilities of success.
4. Useful numerical information may be accessed from computer data banks. Use this data with care and check its quality.
5. Check on the state of your knowledge by attempting to construct a logical framework.

- Judgemental arguments should be expressed in a soundly structured form similar to that suggested by Toulmin (1969).
- A final plea is made for a holistic view of decision making.

References

Allison G. T. (1969). Conceptual models and the Cuban missile crisis. *American Political Science Review,* **LXIII**, 3.

Alvesson M. (1987). *Organization theory and technocratic consciousness.* Walter de Gruyer, Berlin.

Amason A. C. and Schweiger D. M. (1994). Resolving the paradox of conflict, strategic decision making and organizational performance. *The Intnl J. of Conflict Management,* **5**, 3.

Arendt H. (1970). *On violence.* Penguin Press, London.

Armstrong J. S. (1985). *Long-range forecasting; from crystal ball to computer.* Wiley–Interscience, New York.

Armstrong P. (1984). Competition between the organizational professions and the evolution of management control strategies. In Thompson K. (ed.), *Work, employment and unemployment.* Open University Press, Milton Keynes.

Armstrong P. (1987). Engineers, management and trust. *Work, Employment and Society,* **1**, 4, pp. 421–440.

Asch S. E. (1956). Studies of independence and submission to group pressure. *Psychological Monographs,* 70.

Babcock D. L. (1991). *Managing engineering and technology.* Prentice-Hall, London.

Baron J. (1988). *Thinking and deciding.* Cambridge University Press, Cambridge.

Bastide S. *et al.* (1989). Risk perception and social acceptability of technologies: the French case. *Risk Analysis,* **9**, 2.

Bazerman M. H. (1986). *Judgement in managerial decision making*. John Wiley, New York.

Beach L. R. (1990). *Image theory: decision making in personal and organizational contexts*. John Wiley, Chichester.

Beach L. R. (1993). Image theory: personal and organizational decisions. In Klien G. A. *et al.* (eds), *Decision making in action: models and methods*. Ablex, Norwood, New Jersey.

Beck U. (1992). *Risk society: towards a new modernity*. Sage, London.

Berg M. (1994). *The age of manufactures 1700–1820: industry innovation and work in Britain*. Routledge, London.

Bloor G. and Dawson P. (1994). Understanding professional culture in organizational context. *Organization Studies*, **15**, 2.

Bolman L. G. and Deal T. E. (1991). *Reframing organizations*. Jossey-Bass, San Francisco.

Bord R. J. and O'Connor R. E. (1990). Risk communication, knowledge and attitudes: explaining reactions to a technology perceived as risky. *Risk Analysis*, **10**, 4.

Brehmer B. (1980). Probabilistic functionalism in the laboratory: learning and interpersonal (cognitive) conflict. In Hammond K. R. *et al.* (eds), *Realizations of Brunswik's representative design*. Jossey-Bass, San Francisco.

Brehmer B. (1986). The role of judgement in small group conflict and decision-making. In Arkes H. R. and Hammond K. R. (eds), *Judgement and decision making*. Cambridge University Press, Cambridge.

Brown J. (1984). Professional hegemony and analytical possibility: the interaction of engineers and anthropologists in project development. In Millsap, W. (ed.), *Applied social science for environmental planning*. Westview Press, Boulder, Colorado.

Brunswik E. (1952). The conceptual framework of psychology. *Intnl Encyclopedia of Unified Science*, **1**, 10, University of Chicago Press, Chicago.

Burris B. (1989). Technocratic organization and control. *Organization Studies*, **10**, 1.

Buss D. M. and Craik K. H. (1983). Contemporary world views: personal and policy implications. *J. of Applied Social Psychology*, 13.

Buss D. M. *et al.* (1986). Contemporary world views and perceptions of the technological system. In Covello V. T. *et al.* (eds), *Risk evaluation and management*. Plenum Press, New York.

Callon M. (1986). Some elements of a sociology of translation: domestication of the scallops and the fishermen of St Brienc Bay. In Law J. (ed.), *Power, action and belief: a new sociology of knowledge?*. Routledge and Kegan Paul, London.

Callon M. (1987). Society in the making: the study of technology as a tool for sociological analysis. In Bijker W. E. *et al.* (eds), *The social construction of technological systems: new directions in the sociology and history of technology.* MIT Press, Cambridge, Mass.

Callon M. (1991). Techno-economic networks and irreversibility. In Law J. (ed.), *A sociology of monsters: essays on power, technology and domination*. Routledge, London.

Clegg S. R. (1990). *Modern organizations: organization studies in the postmodern world*. Sage, London.

Cohen M. D. *et al.* (1972). A garbage can model of organizational choice. *Administrative Science Quarterly*, **17**, 1.

Coleman G. (1987). Logical framework approach to the monitoring and evaluation of agricultural and rural development projects. *Project Appraisal*, **2**, 4.

Collins R. (1990a). Changing conceptions in the sociology of the professions. In Torstendahl R. and Burrage M. (eds), *The formation of professions*. Sage, London.

Collins R. (1990b). Market closure and the conflict theory of the professions. In Burrage M. and Torstendahl R. (eds), *Professions in theory and history.* Sage, London.

Cooksey R. W. (1995). *Judgement analysis: theory, methods, and applications*. Academic Press, San Diego.

Cox R. A. (1982). Improving risk assessment methods for process plant. *J. of Hazardous Materials*, 6.

Craib I. (1992). *Anthony Giddens*. Routledge, London.

Crespi F. (1992). *Social action and power.* Blackwell, Oxford.

Cyert R. M. and March J. G. (1963). *A behavioral theory of the firm.* Prentice-Hall, Englewood Cliffs, N.J.

Dawes A. (1970). The action frame of reference. In Thompson K. and Tunstall J. (eds), *Sociological perspectives*. Penguin Books, Harmondsworth.

Dean J. W. and Sharfman M. P. (1993). Procedural rationality in the strategic decision making process. *J. of Management Studies*, **30**, 4.

Delbecq A. L. and Gustofson D. H. (1975). *Group techniques for project planning: a guide to nominal group and Delphi processes*. Scott Foresman, Illinois.

Deluca D. R. *et al.* (1986). Public perceptions of technological risks; a methodological study. In Covello V. T. *et al.* (eds), *Risk evaluation and management*. Plenum Press, New York.

Di Maggio P. and Powell W. W. (1983). The iron cage revisited; institutional isomorphism and collective rationality in organizational fields. *American Sociological Review*, 48.

Donaldson L. (1985). *In defence of organization theory: a response to the critics*. Cambridge University Press, Cambridge.

Dryden P. S. and Gawecki L. J. (1987). The transport of hazardous goods: an approach to identifying and apportioning costs. *12th Australian Transport Research Forum*, Brisbane.

Ducot C. and Lubben G. J. (1980). A typology for scenarios. *Futures*, 12.

Eden C. *et al.* (1983). *Messing about in problems*. Pergamon, Oxford.

Eiser J. R. (1990). *Social judgement*. Open University Press, Milton Keynes.

Ellis D. G. and Fisher B. A. (1994). *Small group decision making: communication and the group process*. McGraw-Hill, New York.

European Federation of Chemical Engineers (EFCE) (1985). *Risk analysis in the process industries*. Report of the international study group on risk analysis. Institution of Chemical Engineers, London.

Fast J. C. and Looper L. T. (1988). *Multi-attribute decision modelling techniques: a comparative analysis*. Report No. AFHRL-TR-88-3, Air Force Human Resources Laboratory, Texas.

Fayol H. (1949). *General and industrial management*. Pitman, London.

Fischer F. (1990). *Technocracy and the politics of expertise*. Sage, Newbury Park, California.

Fredrickson J. W. (1986). The strategic decision process and organizational structure. *Academy of Management Review*, **11**, 2.

French S. (1989). *Readings in decision analysis*. Chapman and Hall, London.

Galbraith J. R. (1977). *Organization design*. Addison-Wesley, Reading, MA.

Giddens A. (1984). *The constitution of society*. Polity Press, Cambridge.

Glasbergen P. (ed.) (1995). *Managing environmental disputes: network management as an alternative*. Kluwer Academic, Dordrecht, Netherlands.

Glover I. A. and Kelly M. P. (1987). *Engineers in Britain, a sociological study of the engineering dimension*. Allen and Unwin, London.

Goodwin P. and Wright G. (1991). *Decision analysis for management judgement*. John Wiley, Chichester.

Hales C. P. (1986). What do managers do? A critical review of the evidence. *J. of Management Studies*, **23**, 1.

Hammond K. R. (1993). Naturalistic decision making from a Brunswikian viewpoint: its past, present, future. In Klein G. A. *et al.* (eds), *Decision-making in action: models and methods*. Ablex, Norwood, New Jersey.

Hammond K. R. *et al.* (1975). Social judgement theory. In Kaplan M. F. and Schwartz S. (eds), *Human judgement and decision process: formal and mathematical approaches*. Academic Press, New York.

Hammond K. R. *et al.* (1983). *Direct comparison of intuitive, quasi-rational and analytic cognition*. Rep. No. 248, University of Colorado, Center for Research on Judgement and Policy.

Harrison F. E. and Pelletier M. A. (1993). A typology of strategic choice. *Technological Forecasting and Social Change*, 44.

Hartley J. (1983). Ideology and organizational behaviour. *Intnl Studies of Management and Organizations*, **XIII**, 3.

Hickson D. J. *et al.* (1986). *Top decisions: strategic decision making in organizations*. Basil Blackwell, Oxford.

Hill G. W. (1982). Group versus individual performance: are n + 1 heads better than one? *Psychological Bulletin*, 91.

Hill S. (1988). *The tragedy of technology.* Pluto Press, London.

Hirokawa R. Y. (1992). Communication and group decision-making efficacy. In Cathcart R. S. and Samovar L. A. (eds), *Small group communication*. Brown, Dubuque, IA.

Hofstede G. (1984). Motivation, leadership, and organization; do American theories apply abroad?. In Kolb D. A. *et al.* (eds), *Organizational psychology.* Prentice-Hall, Englewood Cliffs, N.J.

Hofstede G. *et al.* (1990). Measuring organizational cultures: a qualitative and quantitative study across twenty cases. *Administrative Science Quarterly,* 35.

Hogarth R. M. (1987). *Judgement and choice, the psychology of decision.* John Wiley, Chichester.

Hwang C. and Yoon K. (1981). *Multiple attribute decision making: methods and applications*. Springer-Verlag, Berlin.

Hynes M. and Vanmarke E. (1976). Reliability of embankment performance predictions. *Proc. ASCE Engrg Mech. Divisions Speciality Conference*. University of Waterloo Press, Waterloo, Canada.

Isenberg D. J. (1988). How managers think. In Bell D. E. *et al.* (eds), *Decision making: descriptive, normative, and prescriptive interactions*. Cambridge University Press, Cambridge.

Jaccard J. *et al.* (1989). Couple decision making: individual and dyadic level analysis. In Brinberg D. and Jaccard J. (eds), *Dyadic decision making*. Springer-Verlag, New York.

Jackall R. (1988). *Moral mazes: the world of corporate managers*. Oxford University Press, Oxford.

Jamous H. and Peloille B. (1970). Profession or self-perpetuating systems? Changes in the French hospital system. In Jackson J. A. (ed.), *Professions and professionalism*. Cambridge University Press, Cambridge.

Janis I. L. (1972). *Victims of groupthink*. Houghton Mifflin, Boston.

Janis I. L. (1989). *Crucial decisions: leadership in policy and crisis management*. The Free Press, New York.

Janis I. L. and Mann L. (1977). *Decision making*. The Free Press, New York.

Kahneman D. and Tversky A. (1979). Prospect theory: an analysis of decision under risk. *Econometrica*, 47.

Kasperson R. E. *et al.* (1988). The social amplification of risk: a conceptual framework. *Risk Analysis*, **8**, 2.

Kelly G. A. (1955). *The psychology of personal constructs*. Norton, New York.

Kipnis D. and Schmidt S. M. (1983). An influence perspective on bargaining within organizations. In Bazerman M. H. and Lewicki R. J. (eds), *Negotiating in organizations*. Sage, Beverley Hills, California.

Klein G. A. (1993). A recognition-primed decision model of rapid decision making. In Klien G. A. *et al.* (eds), *Decision making in action: models and methods*. Ablex, Norwood, New Jersey.

Kline T. J. B. (1994). Measurement of tactical and strategic decision making. *Educational and Psychological Measurement*, **54**, 3.

Kleindorfer P. R. *et al.* (1993). *Decision sciences: an integrative perspective*. Cambridge University Press, Cambridge.

Korsgaard M. A. *et al.* (1995). Building commitment, attachment, and trust in strategic decision-making teams: the role of procedural justice. *Academy of Management Journal*, **38**, 1.

Kriger M. P. and Barnes L. B. (1992). Organizational decision making as hierarchical levels of drama. *J. of Management Studies*, **29**, 4.

Kumar K. (1986). *Prophesy and progress: The sociology of industrial and post-industrial society*. Penguin Books, Harmondsworth.

Kumar K. (1988). *The rise of modern society*. Basil Blackwell, Oxford.

Kunreuther H. (1992). A conceptual framework for managing low-probability events. In Krimsky S. and Golding D. (eds), *Social theories of risk*. Praeger, Westport, Conn.

Larson M. S. (1977). *The rise of professionalism: a sociological analysis*. University of California Press, Berkeley.

Latour B. (1987). *Science in action: how to follow scientists and engineers through society*. Open University Press, Milton Keynes.

Law J. (1986). On the methods of long-distance control: vessels, navigation and the Portuguese route to India. In Law J. (ed.), *Power, action and belief: a new sociology of knowledge?*. Routledge and Kegan Paul, London.

Law J. (1991). Power, discretion and strategy. In Law J. (ed.), *A sociology of monsters: essays on power, technology and domination*. Routledge, London.

Law J. (1992a). The Olympus 320 engine: a case study in design, development, and organizational control. *Technology and Culture*, 33.

Law J. (1992b). Notes on the theory of the actor-network: ordering, strategy and heterogeneity. *Systems Practice*, **5**, 4.

Law J. (1994). *Organizing modernity*. Blackwell, Oxford.

Law J. and Callon M. (1992). The life and death of an aircraft: a network analysis of technical change. In Bijker W. E. and Law J. (eds), *Shaping technology/building society: studies in sociotechnical change*. MIT Press, Cambridge, Mass.

Lichtenstein S. *et al.* (1982). Calibration of probabilities: the state of the art to 1980. In Kahneman D. *et al.* (eds), *Judgement under uncertainty: heuristics and biases*. Cambridge University Press, Cambridge.

Lindblom C. E. (1959). The science of muddling through. *Public Administration Review*, 19.

Lindblom C. E. (1979). Still muddling, not yet through. *Public Administration Review*, 39.

Linstone H. A. (1984). *Multiple perspectives for decision making: bridging the gap between analysis and action*. North-Holland, New York.

Lloyd B. U. (1991). *Engineers in Australia, a profession in transition*. Macmillan, Melbourne.

Lukes S. (1974). *Power: a radical view*. Macmillan, London.

Luthans F. R. *et al.* (1988). *Real managers*. Ballinger, Cambridge, MA.

March J. G. (1982). Theories of choice and making decisions. *Transactions: Social Science and Modern Society*, 20.

March J. G. (1994). *A primer on decision making: how decisions happen*. The Free Press, New York.

March J. G. and Sevon G. (1988). Gossip, information and decision-making. In March J. G. (ed.), *Decisions and organizations*. Basil Blackwell, Oxford.

March J. G. and Shapira Z. (1988). Managerial perspectives on risk and risk-taking. In March J. G. (ed.), *Decisions and organizations*. Basil Blackwell, Oxford.

March J. G. and Shapira Z. (1992). Behavioral decision theory and organizational decision theory. In Zey M. (ed.), *Decision making: alternatives to rational choice models*. Sage, London.

March J. G. and Simon H. A. (1958). *Organizations*. John Wiley, New York.

Mayer J. P. (1956). *Max Weber and German politics*. Faber and Faber, London.

McCall M. W. and Kaplan R. E. (1985). *Whatever it takes: decision makers at work*. Prentice-Hall, Englewood Cliffs, N.J.

Mintzberg H. (1973). *The nature of managerial work*. Harper and Row, New York.

Mintzberg H. (1984). Planning on the left side and managing on the right. In Kolb D. A. *et al.* (eds), *Organizational psychology*. Prentice-Hall, Englewood Cliffs, N.J.

Mintzberg H. *et al.* (1976). The structure of unstructured decision processes. *Administrative Science Quarterly*, 21.

Mitchell T. R. *et al.* (1988). *People in organizations: an introduction to organizational behaviour in Australia*. McGraw-Hill, Sydney.

Molloy S. and Schwenk C. R. (1995). The effects of information technology on strategic decision making. *J. of Management Studies*, **32**, 3.

Montgomery H. (1993). The search for a dominance structure in decision making: examining the evidence. In Klein G. A. *et al.* (eds), *Decision making in action: models and methods*. Ablex, Norwood, New Jersey.

Moore P. G. and Thomas H. (1976). *The anatomy of decisions*. Penguin Books, Harmondsworth.

Morgan M. G. (1986). *Images of organization*. Sage, Newbury Park, California.

Moscovici S. and Zavalloni M. (1969). The group as a polarizer of attitudes. *J. of Personality and Social Psychology*, 12.

Moscovici S. *et al.* (1969). Influence of a consistent minority on the response of a majority in a color perception task. *Sociometry*, 32.

Mullen J. D. and Roth B. M. (1991). *Decision making: its logic and practice*. Rowan and Littlefield, Savage, Maryland.

NATO ASI Series (1984). *Technological risk assessment*. Martinus Nijhoff Publishers, The Hague.

Nohria N. (1992). Is a network perspective a useful way to studying organizations? In Nohria N. and Eccles R. G. (eds), *Networks and organizations: structure, form, and action*. Harvard Business School Press, Boston, MA.

Nohria N. and Eccles R. G. (eds) (1992). *Networks and organizations: structure, form, and action*. Harvard Business School Press, Boston, MA.

Nutt P. C. (1993a). Flexible decision styles and the choices of top executives. *J. of Management Studies*, 30.

Nutt P. C. (1993b). The identification of solution ideas during organizational decision making. *Management Studies*, **39**, 9.

Orasanu J. and Salas E. (1993). Team decision making in complex environments. In Klein G. A. *et al.* (eds), *Decision making in action: models and methods*. Ablex, Norwood, New Jersey.

Otway H. J. and von Winterfeldt D. (1982). Beyond acceptable risk: on the social acceptability of technologies. *Policy Sciences*, 14.

Parkin J. V. (1993). *Judging plans and projects: analysis and public participation in the evaluation process*. Avebury, Aldershot.

Parkin J. V. (1994a). A power model of urban infrastructure decision making. *Geoforum*, **25**, 2.

Parkin J. V. (1994b). *Public management: technocracy, democracy and organizational reform*. Avebury, Aldershot.

Parkin J. V. (1996). Organizational decision making and the project manager. *Intnl J. of Project Management*, (forthcoming).

Parnaby J. (1985). in *The Engineer*, 5 Dec., pp. 43–5.

Parsons T. (1956). Suggestions for sociological approach to the theory of organizations. *Administrative Science Quarterly*, **1**, pp. 63–65, 225–39.

Parsons T. (1963). On the concept of political power. *Proc. American Philosophical Society*, 107.

Payne J. W. *et al.* (1993). *The adaptive decision maker*. Cambridge University Press, Cambridge.

Penrose E. (1980). *The theory of the growth of the firm*. Blackwell, Oxford.

Peters T. J. and Waterman R. H. (1982). *In search of excellence*. Harper and Row, New York.

Pfeffer J. (1978). *Organization design*. AHM, Arlington Heights, IL.

Philipson L. L. and Napadensky H. S. (1982). The methodologies of hazardous materials transportation risk assessment. *J. of Hazardous Materials*, 6.

Pitblado R. M. (1984). *Hazards study of liquified petroleum gas in automotive retail outlets*. Department of Environment and Planning, Sydney.

Plous S. (1993). *The psychology of judgement and decision making*. McGraw-Hill, New York.

Raelin J. (1985). *The clash of cultures: managers and professionals*. Harvard Business School Press, Boston, MA.

Reader W. J. (1966) *Professional men*. Weidenfeld and Nicolson, London.

Reason J. (1990). *Human error*. Cambridge University Press, Cambridge.

Robson M. (1993). *Problem solving in groups*. Gower, Aldershot.

Rowe A. J. and Boulgarides J. D. (1992). *Managerial decision making*. Macmillan, New York.

Saaty T. L. (1980). *The analytic hierarchy process*. McGraw-Hill, New York.

Schoemaker P. J. H. (1991). *When and how to use scenario planning: a heuristic approach with illustration*. John Wiley, Chichester.

Silverman D. (1970). *The theory of organizations: a sociological framework*. Heinemann, London.

Simon H. A. (1947). *Administrative behaviour*. Macmillan, New York.

Simon H. A. (1965). On the concept of organizational goal. *Administrative Science Quarterly*, 9.

Simon H. A. (1976). *Administrative behaviour: a study of decision-making processes in administrative organization*. Free Press, New York.

Simon H. A. (1986). Alternative visions of rationality. In Arkes H. R. and Hammond K. R. (eds), *Judgment and decision making*.

Cambridge University Press, Cambridge.

Simon H. A. and Associates. (1992). Decision making and problem solving. In Zey M. (ed.), *Decision making: alternatives to rational choice models*. Sage, London.

Sinclair A. (1992). The tyranny of a team ideology. *Organization Studies*, **13**, 4.

Slovic P. (1992). Perceptions of risk: reflections on the psychometric paradigm. In Krimsky S. and Golding D. (eds), *Social theories of risk*. Praeger, Westport, Conn.

Slovic P. *et al.* (1982). Facts versus fears: understanding perceived risk. In Kahneman D. *et al.* (eds), *Judgment under uncertainty: heuristics and biases*. Cambridge University Press, Cambridge.

Slovic P. *et al.* (1986). The psychometric study of risk perception. In Covello V. T. *et al.* (eds), *Risk evaluation and management*. Plenum Press, New York.

Stallen P. J. M. and Thomas A. (1984). Psychological aspects of risk: the assessment of threat and control. In NATO ASI Series, *Technological risk assessment*, Martinus Nijhoff Publishers, The Hague.

Stohl C. and Redding W. C. (1987). Messages and message exchange processes. In Jablin F. M. *et al.* (eds), *Handbook of organizational communication: an interdisciplinary perspective*. Sage, Newbury Park, California.

Stoner J. A. F. (1968). Risky and cautious shifts in group decision: the influence of widely held values. *J. of Experimental Social Psychology*, 4.

Sundstrom E. *et al.* (1990). Work teams: applications and effectiveness. *American Psychologist*, **45**, 2.

Thomas K. W. (1977). Towards multi-dimensional values in teaching: the example of conflict behaviours. *Academy of Management Review*, 12.

Thompson E. P. (1980). *The making of the English working class*. Penguin Books, Harmondsworth.

Toulmin S. E. (1969). *The uses of argument*. Cambridge University Press, Cambridge.

Toulmin S. E. *et al.* (1979). *An introduction to reasoning*. Macmillan, New York.

Trainer F. E. (1982). *Dimensions of moral thought*. New South Wales University Press, Sydney.

Tversky A. (1972). Elimination of aspects: a theory of choice. *Psychological Review*, 79.

Tversky A. and Kahneman D. (1982). Judgement under uncertainty: heuristics and biases. In Kahneman D. *et al.* (eds), *Judgment under uncertainty: heuristics and biases*. Cambridge University Press, Cambridge.

Tversky A. and Kahneman D. (1985). The framing of decisions and the psychology of choice. In Wright G. (ed.), *Behavioural decision making*. Plenum Press, New York.

von Winterfeldt D. and Edwards W. (1986). *Decision analysis and behavioural research*. Cambridge University Press, Cambridge.

von Winterfeldt D. *et al.* (1981). Cognitive components of risk ratings. *Risk Analysis*, **1**, 4.

Vroom V. H. and Yetton P. W. (1973). *Leadership and decision making*. University of Pittsburgh Press, Pittsburgh.

Wartofsky M. W. (1986). Risk, relativism, and rationality. In Covello V. T. *et al.* (eds), *Risk evaluation and management*. Plenum Press, New York.

Watson S. R. and Buede D. M. (1987). *Decision synthesis: the principles and practice of decision analysis*. Cambridge University Press, Cambridge.

Weber M. (1948). From *Max Weber: essays in sociology*, translated by Gerth H. H. and Mills C. W. Routledge and Kegan Paul, London.

Weber M. (1971). Power and bureaucracy. In Thompson K. and Tunstall J. (eds), *Sociological perspectives*. Penguin Books, Harmondsworth.

Whitley R. (1989). On the nature of managerial tasks and skills: their distinguishing characteristics and organization. *J. of Management Studies*, **26**, 3.

Whittaker J. (1991). A reappraisal of probabilistic risk analysis. *Engineering Management Journal*, **3**, 3.

Willmott H. (1990). Beyond paradigmatic closure in organizational enquiry. In Hassard J. and Pym D. (eds), *The theory and philosophy of organizations*. Routledge, London.

Wright G. N. (1984). *Behavioural decision theory: an introduction*. Penguin Books, Harmondsworth.

Wright G. N. and Phillips L. D. (1980). Cultural variation in probabilistic thinking: alternative ways of looking at uncertainty. *Intnl J. of Psychology,* 15.

Zey M. (1992). Criticisms of rational choice models. In Zey M. (ed.), *Decision making: alternatives to rational choice models*. Sage, Newbury Park, California.

Index of sources and main topics (first citation)